GETTING STARTED
WITH
MATHEMATICA®

C-K. Cheung
G. E. Keough
Charles Landraitis
Robert H. Gross

Boston College

John Wiley & Sons, Inc.

New York • Chichester • Weinheim • Brisbane • Singapore • Toronto

Cover image by Marjory Dressler

ISBN 0-471-24050-8

Printed in the United States of America

10 9 8 7 6 5 4 3

Printed and bound by Hamilton Printing Company
Cover printed by The Lehigh Press, Inc.

Preface

Using this Guide

This text is written to enable you to take rapid advantage of the powerful *Mathematica* system in your work or studies, even if you have little or no relevant experience. Both versions 2.0 and 3.0 of the software are covered in this text. Chapters are organized to speed you through basic information and on to your areas of interest. Easy-to-follow explanations are liberally supplemented with illustrative examples, useful tips and trouble-shooting advice. With these features, we are confident that your learning experience will be vivid, pleasurable and, as far as possible, frustration-free.

This text is designed as a tutorial with emphasis on getting you quickly "up to speed." It can also be used as a supplement or reference for students taking a mathematics or science course that requires use of *Mathematica*, such as Calculus, Multivariable Calculus, Advanced Calculus, Linear Algebra, Discrete Mathematics, Modeling, or Statistics.

About *Mathematica*

Mathematica is computer algebra software developed by Wolfram Research, Incorporated, which lets you use the computer like an interactive mathematics scratchpad. *Mathematica* can perform abstract computation as well as numerical computation. It can draw graphics and let you write programs as well. It is a useful tool not only for an undergraduate mathematics or science major, but also for graduate students and researchers. The program is widely used as well by engineers, physicists, transportation officials, and architects.

Organization of the Guide

The Guide is organized as follows:

- Chapter 1 gives a short demonstration of what you'll see in the remaining parts of the Guide. Chapters 2 through 10 contain the basic information that almost every user of *Mathematica* should know.

- Chapters 11 through 14 demonstrate *Mathematica*'s capabilities in single-variable calculus. This includes working with limits, derivatives, integrals, series and differential equations.

- Chapters 15 through 20 cover multivariable calculus and linear algebra. Here you'll find discussion of partial derivatives, multiple integrals, vectors, vector fields and line and surface integrals.

- Chapters 21 and 22 introduce the statistical capabilities of *Mathematica*.

- Chapters 23 through 26 address a collection of topics ranging from animation and simulation to programming and list processing.

- An Appendix explains how to work in the *Mathematica* notebook environment.

Chapter Structure

Each chapter of the Guide has been structured around an area of undergraduate mathematics. Each quickly defines relevant commands, addresses their syntax, and provides basic examples.

Every chapter ends with as many as three special sections that can be passed over during your first reading. However, these sections will provide valuable support when you start asking questions and looking for more detail. These three sections are:

- "More Examples." Here, you'll find more technical examples or items that address more mathematical points.
- "Useful Tips." This section contains some simple pointers that all *Mathematica* users eventually learn. We've drawn them from our experiences in teaching undergraduates how to use *Mathematica*.
- "Troubleshooting Q & A." Here we present a question and answer dialogue on common problems and error messages. You'll also be able to find out more about *how* certain commands work. Many useful topics are covered here.

Conventions Used in This Guide

Almost every *Mathematica* input and output you see in this Guide appears exactly as we executed it in a *Mathematica* session. Slight changes were made only to enhance page layout or to guarantee better photo-offset production quality for graphics.

The entries in our sections on "Useful Tips" are rated with *light bulbs* ♀ . More light bulbs indicate tips that we think are more important than others. Our scale is 1 to 4 light bulbs, but your wattage may vary.

Final Comments

First, let us thank all of our colleagues and students at Boston College who have contributed to the completion of this Guide. Special thanks are extended to Jenny Baglivo, Nancy Gaff, and Sarah Quebec for their contributions to the statistics pages of the Guide. Bill Keane contributed several suggestions on the penultimate printing of the Guide. Bill Zahner was extremely helpful for his reading, rereading, and critiquing of (many versions of) the Guide as it developed. He also offered terrific suggestions on presentation and arrangement.

The staff at John Wiley, Inc., have been very supportive of our efforts. Our sincerest thanks go to Sharon Smith for her enthusiastic response to this project.

Finally, despite our best efforts, it is likely that somewhere in these many pages, an error of either omission or commission awaits you. We sincerely apologize if this is the case and accept full responsibility for any inaccuracies.

We will make available on the World Wide Web a listing of further comments, examples and any inaccuracies that may be found in this Guide. You can locate this information by navigating the main Boston College website hierarchy down to the Department of Mathematics page. Begin your search at:

<http://www.bc.edu>

We will also try to answer any inquiries you may have about the information presented in the Guide. Feel free to share your teaching/learning experience with us. You can reach us through the following e-mail addresses:

ck.cheung@bc.edu, keough@bc.edu
charles.landraitis@bc.edu, rob.gross@bc.edu

We wish you only the best computing experiences in *Mathematica*.

**C-K, Jerry,
Charlie, and Rob**

March, 1998

Contents

Part 1. Basic *Mathematica* Commands

1. Running *Mathematica*   1

Computer Systems 1
A Quick Tour 2
More Examples 6
Useful Tips 7
Troubleshooting Q & A 7

2. Calculator Features in *Mathematica*   8

Simple Arithmetic 8
Output Styles 10
Built-in Constants and Functions 10
Error Messages 12
More Examples 12
Useful Tips 13
Troubleshooting Q & A 13

3. Variables and Functions   15

Variables 15
Functions 17
More Examples 18
Useful Tips 19
Troubleshooting Q & A 20

4. Computer Algebra   21

Working with Polynomials
and Powers 21
Working with Rational
Functions 22
Working with Trig and
Hyperbolic Functions 23
Useful Tips 24
Troubleshooting Q & A 24

5. Working with Equations   26

Equations and Their Solutions 26
Equations and Numerical
Solutions 28
FindRoot and Numerical
Solutions 28
More Examples 29
Useful Tips 30
Troubleshooting Q & A 30

6. Lists and Tables   33

Lists 33
Tables 34
Plotting Points 35
More Examples 36
Troubleshooting Q & A 37

7. Getting Help and Loading Packages   38

Text-based Help 38
Browser-based Help 39
Packages 41
Troubleshooting Q & A 42

Part 2.　Drawing Pictures in *Mathematica*

8. Making Graphs　　　　　　　　　**43**

Drawing the Graph of a Function 43
Plotting Multiple Expressions 45
Style Options for Graphics 46

More Examples 48
Useful Tips 49
Troubleshooting Q & A 49

9. Plotting a Curve　　　　　　　　　**51**

Parametric Plots 51

Troubleshooting Q & A 53

10. PolarPlot and ImplicitPlot　　　　**55**

Plotting in Polar Coordinates 55
Plotting Graphs of Equations 56

Useful Tips 58
Troubleshooting Q & A 58

Part 3.　*Mathematica* for One Variable Calculus

11. Limits and Derivatives　　　　　　**60**

Limits 60
Differentiation 61
More Examples 63

Useful Tips 65
Troubleshooting Q & A 65

12. Integration　　　　　　　　　　　**67**

Anti-Differentiation 67
Numerical Integration 68
More Examples 69

Useful Tips 70
Troubleshooting Q & A 71

13. Series, Taylor Series, and Fourier Series　　**72**

Series 72
Taylor Series 73
Fourier Series 75

More Examples 75
Troubleshooting Q & A 77

14. Solving Differential Equations　　　**79**

Symbolic Solutions of Equations 79
Numerical Solutions of Equations ... 80
Systems of Differential Equations ... 81

More Examples 82
Troubleshooting Q & A 83

Part 4.　*Mathematica* for Multivariable Calculus

15. Making Graphs in Space　　　　　**85**

Graphing Functions of
　Two Variables 85
ViewPoints for 3-D Graphics 87

Surfaces in Cylindrical and
　Spherical Coordinates 89

More Examples 91
Helpful Tips 91
Troubleshooting Q & A 92

16. Level Curves and Level Surfaces   94

Level Curves in the Plane 94
Level Surfaces in Space 96
More Examples 96
Troubleshooting Q & A 99

17. Partial Differentiation and Multiple Integration   100

Partial Derivatives 100
Double and Triple Integrals 100
More Examples 102
Troubleshooting Q & A 105

Part 5. *Mathematica* for Linear Algebra and Vector Calculus

18. Matrices and Vectors   106

Vectors 106
Matrices 107
Elementary Row Transformations . 109
Troubleshooting Q & A 110

19. Parametric Curves and Surfaces in Space   112

ParametricPlot3D 112
Plotting Multiple Curves
 and Surfaces 114
Styling Curves and Surfaces 115
More Examples 116
Troubleshooting Q & A 117

20. Vector Fields   119

Drawing a Vector Field 119
Gradient, Curl, and Divergence 121
Line and Surface Integrals 122
More Examples 123
Useful Tips 124
Troubleshooting Q & A 124

Part 6. Using *Mathematica* in Statistics

21. Basic Statistics   126

Graphical Presentation of Data 126
Numerical Measures 127
Probability Distributions 128
More Examples 129
Useful Tips 131
Troubleshooting Q & A 131

22. Regression and Interpolation   133

Regression 133
Interpolation 135
More Examples 135
Useful Tips 139
Troubleshooting Q & A 139

Part 7. Advanced Features of *Mathematica*

23. Animation   140

Getting Started 140
More Examples 142
Useful Tips 145
Troubleshooting Q & A 145

24. More About Lists   147

Displaying Lists 147   More Examples 150
Useful List Commands.................... 148

25. Random Numbers and Simulation   154

Random Numbers 154   Examples in Simulation 155

26. *Mathematica* for Programmers   160

Elements of Traditional
  Programming Languages 160   Pattern Matching 165
Functional Programming................ 163   File I/O 168

Appendix. Working with Notebooks   170

Front End and Kernel 170   Notebook Structure 173
Notebooks and Cells 170   Input Shortcuts for Version 3.0...... 176

Index

CHAPTER 1

Running *Mathematica*

Computer Systems

What Computer System Are You Using?

Mathematica software runs on almost every major computer system including mainframes and desktop systems. *Mathematica* can also be set up to run across a network and even between systems.

Implementations of *Mathematica* generally break down into one of two types:

- Text-based systems (e.g., UNIX or other mainframe systems, and DOS-based personal computers) where you can only type input through a keyboard one line at a time.

- Graphically-based systems (e.g., desktop systems running MS Windows, Solaris, or the MacOS) where you can both type on a keyboard and use a mouse to navigate a window.

The *Mathematica* commands we discuss in this guide will work on both types of systems, although some features discussed in certain chapters may apply only to graphically-based systems.

Starting the Software

You need to follow the instructions that came with the *Mathematica* software to install it on your computer system. Once you've completed the installation, you are ready to explore *Mathematica*.

Starting *Mathematica* obviously depends on the system you are using:

- On text-based systems, you'll enter one of the commands **math** or **mathremote**, depending on how the software has been installed and how the system is configured. The current version of *Mathematica* will be launched, and you will see the line:

```
In[1] :=
```

- On graphically-based systems, you'll typically find the icon of the *Mathematica* application in a window. Click (or double-click) on the icon. The current version of *Mathematica* will be launched, and an empty window will be opened, waiting for your input.

(If you run *Mathematica* over a network, you may need to check with your system manager for the starting procedure.)

If you get an error message when starting *Mathematica*, check the Q & A section at the end of this chapter.

The Ins and Outs

Mathematica is interactive software. For almost every entry you make, *Mathematica* will provide a direct response. Once you launch *Mathematica*, you can type, for example, the expression:

1 + 1

and then press the evaluation key or keys (e.g., the **Enter** key on a Macintosh, or **Shift/Enter** on a PC). Your input will be labeled. Then after a few seconds, *Mathematica* will give you the response:

In[1] := **1 + 1**

Out[1] := 2

The labels *In[1]* and *Out[1]* are supplied automatically by *Mathematica* – you should not enter them yourself. (If you get an error message here, check the Q & A section.)

Now you enter, say,

39 - 11

then press **Enter** or **Shift/Enter** and you get:

In[2] := **39 - 11**

Out[2] := 28

Mathematica numbers every input it receives, as well as every output it produces, starting with 1, from the beginning of a session.

Throughout the rest of this manual, we will not show you the leading labels *In* and *Out*. Our input will be shown in boldface, and the response in plain text. So, in the future, you'll see us write the evaluations above as:

1 + 1

2

39 - 11

28

Quit

If you've had enough and want to exit *Mathematica* from a text-based system, you can simply enter:

Quit[]

On graphically-based systems, you finish a *Mathematica* session by choosing the "Quit" item in the File menu.

A Quick Tour

For the rest of this chapter, we'll show you some of the capabilities of *Mathematica*. We present these examples only to whet your appetite. You can follow along at your computer by typing what we show below. In later chapters, we'll give you a more complete explanation about how to use these commands.

Note: When you input the following commands in *Mathematica*, make sure you do the following:

• Use upper- and lower- case characters exactly as we do. *Mathematica* is very "case sensitive." If you use the wrong capitalization, you can hurt *Mathematica*'s "feelings."

- Use exactly the type of brackets we show. There are three types of them: [*square brackets*], (*parentheses*) and { *curly braces* }. Each has its own meaning in *Mathematica*. If you use the wrong one, *Mathematica* will be confused.

- Whenever you type a command that is more than one line, you move to a second line by pressing the **Return** key. You should break the line exactly as we show in the text.

- Press the "evaluation key" (e.g., the **Enter** key on a Macintosh or the **Shift/Enter** on a PC) to see the output.

Calculator

Mathematica does all the work of a hand-held electronic calculator. You can enter numerical expressions and *Mathematica* will do the arithmetic:

235.567*441.235/623.45

166.718

Sin[0.3]

0.29552

But *Mathematica* can go much further. Try this factorial computation!

321 !

```
679269174457380047028785170185919186947307915378873794717507\
  48348000566996201075565883634067117697871971951788620081\
  90897833975117872915098411594472966982434784667390565661\
  25534997069369223181107508369736925738136722506332041830\
  92581043853551806637709746119945430430888080911106503450571\
  07426224932943371803396277440074116196619232116926339614\
  28696341204992520108400256503261237155571285404597604616\
  47357620275685214063161701206402885960985439459427543149\
  41465184526566990650415696495063334653541359881356653476\
  71738544347224622640956514758437414180328510235242923530\
  79206058818535424000000000000000000000000000000000000000\
  0000000000000000000000000000000000000000
```

You can't get this on your calculator!

Solving Equations

Mathematica can solve complicated equations and even systems of equations in many variables. For example, the equations $2x + 5y = 37$ and $x - 3y = 21$ have a simultaneous solution:

Solve[{ 2x+5y==37, x-3y==21 }, { x, y }]

$$\{\{x \to \frac{216}{11}, y \to -\frac{5}{11}\}\}$$

Mathematica can also find solutions to equations numerically. For example, the equation $x = \cos(x)$ has a solution very close to $x = 0.75$. We can find it with:

FindRoot[x==Cos[x], {x, 0.75}]

$\{x \to 0.739085\}$

Computer Algebra

Mathematica is very good at algebra. It can work with polynomials:

```
Expand[(x-2)^2 * (x+5)^3]
```

$$500 - 200\,x - 115\,x^2 + 19\,x^3 + 11\,x^4 + x^5$$

```
Factor[500 - 200x - 115x^2 + 19x^3 + 11x^4 + x^5]
```

$$(-2 + x)^2\,(5 + x)^3$$

Are you impressed? *Mathematica* also knows standard trigonometric identities such as $\sin^2(x) + \cos^2(x) = 1$:

```
Simplify[Sin[x]^2 + Cos[x]^2]
```

$$1$$

That one was easy, but you probably forgot that $\sec^2(x) - \tan^2(x) = 1$:

```
Simplify[Sec[x]^2 - Tan[x]^2]
```

$$1$$

Calculus

Mathematica even knows a lot about calculus! We can find the derivative of the function $f(x) = x/(1 + x^2)$ with:

```
D[ x/(1+x^2), x]
```

$$-\frac{2\,x^2}{(1 + x^2)^2} + \frac{1}{1 + x^2}$$

A complicated integral such as $\int \frac{1}{1 + x^3}\,dx$ is handled rather easily.

```
Integrate[1/(1+x^3),x]
```

$$\frac{\text{ArcTan}\left[\frac{-1+2x}{\sqrt{3}}\right]}{\sqrt{3}} + \frac{1}{3}\,\text{Log}[1 + x] - \frac{1}{6}\,\text{Log}[1 - x + x^2]$$

Graphing in the Plane

Mathematica does everything a standard graphing calculator does and does it better! For example, to see the graph of the function $f(x) = x/(1 + x^2)$ over the interval $-4 \le x \le 4$, you can use:

```
Plot[ x/(1+x^2), {x, -4, 4} ]
```

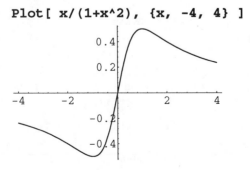

We can see a "daisy" with:

```
ParametricPlot[ {Cos[21*t]*Cos[t],Cos[21*t]*Sin[t]},
      {t,0,2Pi}]
```

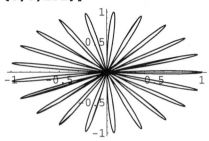

Plotting in Space

Mathematica does a wonderful job with three-dimensional graphics. Let us show you some pictures.

```
Plot3D[Sin[x]*Cos[y],{x,0,2Pi},{y,0,2Pi}]
```

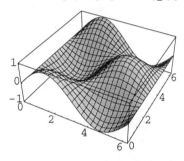

```
ParametricPlot3D[{t/5,  r*Cos[t],  r*Sin[t]},
          {r,  0,  1},  {t,  0,  6Pi} ,
          PlotPoints->{8,60},  ViewPoint->{2,2,1}]
```

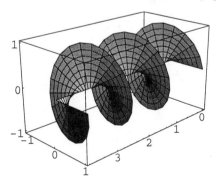

Programming

You can program in *Mathematica* much as you do in a programming language such as Pascal or C. Here's a simple routine to simulate the flipping of a coin several times and return the number of heads observed:

```
coinFlips[ howMany_ ] := Module[
    {t, heads = 0, n = howMany},
    While[ n>0,
        t = Random[];
        If[ t<0.5, heads += 1 ];
        n--;
        ];
    Return[heads]
    ]
```

You can use this routine with statements such as:

```
Print[coinFlips[100]," heads seen in 100 trials"]
```

49 heads seen in 100 trials

```
Print[coinFlips[1000]," heads seen in 1000 trials"]
```

517 heads seen in 1000 trials

More Examples

In the "More Examples" sections of this guide, we present examples involving more mathematics. Students of mathematics, science, and engineering may find these of interest.

Here's an example. Not too many people know about Bessel functions. But if you're learning physics or engineering, you might want to know what *Mathematica* has available for you in Bessel functions.

Special Functions

The Bessel function of order 0, $J_0(x)$, is a solution to the differential equation:

$$x^2 y'' + xy' + x^2 y = 0.$$

You can see its graph with:

```
Plot[ BesselJ[0,x], {x,0,20} ]
```

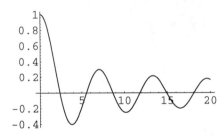

The smallest, positive zero of $J_0(x)$ is at approximately $x = 2.40483$:

```
FindRoot[ BesselJ[0,x]==0, {x, 2.0} ]
```

$\{x \to 2.40483\}$

Useful Tips

 If you are running *Mathematica* on a personal computer, it is better to quit other applications before you start *Mathematica*. If you need to run another program at the same time as you run *Mathematica*, start with *Mathematica* first.

Troubleshooting Q & A

Most chapters of this guide end with a Troubleshooting section, where we answer some common questions we think you'll have. But there's not too much you can ask yet, except:

Question ... When I tried to start *Mathematica*, I got an error message that there wasn't enough memory. What should I do?

Answer ... *Mathematica* is memory intensive. Make sure that you have enough RAM to run the software. Also, it is a good practice to quit other applications before you launch *Mathematica*.

Question ... When I tried to enter the first input, *Mathematica* gave me an error message that it cannot establish a link to a kernel. What went wrong?

Answer ... If you are running *Mathematica* over a network, check with your system manager. If you are working on a personal computer, *Mathematica* should set up an automatic "local link" during its installation. Check your configuration by referring to your *Mathematica* manual.

Question ... What do I do now?

Answer ... Turn the page and start learning about *Mathematica*! Also, if you are using a graphically-based system, check the Appendix for more information on how to work with Notebooks.

CHAPTER 2

Calculator Features in *Mathematica*

Simple Arithmetic

Basic Arithmetic Operations

In this chapter, we will show you how *Mathematica* can work as a calculator. We start with basic arithmetic.

The basic arithmetic operations of *Mathematica* are:

Command	What it does
+, −	Add, Subtract
*, /	Multiply, Divide
^	Raise to a power (exponentiation)

Let's calculate: $\dfrac{25.5}{5}$, $4 + 2^5$, and $\dfrac{23}{5} - \dfrac{3}{5} + 5(2^3)$:

25.5/5

5.1

4+2^5

36

23/5 - 3/5 + 5*2^3

44

We can also use parentheses to group terms. For example, $(3+4)\left(\dfrac{4-8}{5}\right)$ is entered with:

(3+4)*((4−8)/5)

$$-\frac{28}{5}$$

Note: Use (*parentheses*) to group terms in expressions. Do not use [*square brackets*] or { *curly braces* } – they mean something different.

Precedence

As you noticed above, *Mathematica* follows the laws of precedence of multiplication over addition, and so on, just as you do by hand.

For example,

9/6*22+5

38

actually computes $(\frac{9}{6} \times 22) + 5$. A common mistake is to think of the input as either $\frac{9}{6 \times 22} + 5$ or $\frac{9}{6} \times (22 + 5)$.

Spaces Mean Multiplication

If you leave space between terms, they get multiplied together (but this is not good programming style!)

27 3

81

Comments

You can add a comment to any expression by enclosing it within the symbol pairs (* and *). For example:

27 3   (* This multiplies 27 and 3. *)

81

Mathematica will neglect the phrase "**This multiplies 27 and 3.**" when it evaluates your input. The phrase is just for your own reference.

We'll use comments like this throughout this guide to remind you what we're emphasizing.

Previous Results and % Syntax

As a convenience, *Mathematica* lets you use the percent sign "%" to stand for "the last result obtained," so you can avoid retyping an expression.

For example,

3200*12

38400

% - 6500   (* Same as 38400 - 6500.*)

31900

(1000 + %)^2   (* Same as (1000+31900)^2. *)

1082410000

You can use the double percent sign "%%" for the "second-last result obtained," a triple percent sign "%%%" for the third-last, and so on. You can also refer to specific outputs by number with the percent sign. For example, **%16** is shorthand notation for Out[16].

Note: Using the percent sign % can be confusing. We will use it sparingly.

Output Styles

Calculator-Style Values

Mathematica normally gives you an exact (symbolic) value for every expression:

```
(3+9) * (4-8) / 1247 * 67
```

$$-\frac{3216}{1247}$$

You can force *Mathematica* to give you an answer that looks like the decimal answer you'd get on a calculator by using **N** with square brackets around an expression:

```
N[(3+9) * (4-8) / 1247 * 67]
```
```
-2.57899
```

An alternative syntax is to use **// N** after the input (this is called **postfix syntax**)

```
(3+9) * (4-8) / 1247 * 67 // N
```
```
-2.57899
```

You can see more digits in the answer – say 40 – with:

```
N[(3+9)*(4-8)/1247*67, 40]    (*The ,40 makes N
                                  give 40 digits. *)
```
```
-2.5789895749799518845228548516439455469126
```

Results like these are called **approximate numeric values** in *Mathematica*.

Scientific Notation

Mathematica uses standard scientific notation to display results when the numbers either get very large or very small:

```
N[1234567890]
```

$$1.23457 \times 10^9$$

```
0.000003492836
```

$$3.492836 \times 10^{-6}$$

Built-in Constants and Functions

Built-in Constants

The mathematical constants used most often are already built into *Mathematica*. Notice that all begin with a capital letter:

Constant	Value	Explanation	Mathematica
π	3.1415926...	Ratio of a circle's circumference to its diameter	**Pi**
e	2.71828...	Natural Exponential	**E**
i	$i = \sqrt{-1}$	Imaginary Number	**I**

$\frac{\pi}{180}$	.0174532...	Degree to radian conversion multiplier	**Degree**
∞	∞	(positive) infinity	**Infinity**

For example, to see the value of π to 45 significant digits, we use:

N[Pi,45]

3.14159265358979323846264338327950288419716940

To compute the numerical value of $\pi^4 - 5e^{1/3}$, we type:

N[Pi^4 - 5*E^(1/3)]

90.431

Built-in Functions

Mathematica has many built-in functions. These are the functions you'll probably use the most.

Function(s)	*Sample(s)*	Mathematica *Name(s)*
Natural logarithm	$\ln(x)$	**Log[x]**
Logarithm to base a	$\log_a x$	**Log[a, x]**
Exponential	e^x	**Exp[x]**
Absolute value	$\lvert x \rvert$	**Abs[x]**
Square root	$\sqrt{x}$	**Sqrt[x]**
Trigonometric	$\sin(x)$, $\cos(x)$, ...	**Sin[x], Cos[x], Tan[x], Cot[x], Sec[x], Csc[x]**
Inverse trigonometric	$\sin^{-1}(x)$, $\cos^{-1}(x)$, ...	**ArcSin[x], ArcCos[x], ArcTan[x], ArcCot[x]**, etc.
Hyperbolic	$\sinh(x)$, $\cosh(x)$, ...	**Sinh[x], Cosh[x], Tanh[x], Coth[x], Sech[x], Csch[x]**
Inverse hyperbolic	$\sinh^{-1}(x)$, $\cosh^{-1}(x)$, ...	**ArcSinh[x], ArcCosh[x], ArcTanh[x], ArcCoth[x]**, etc.

For example,

Sin[Pi]

0

N[Sin[180]]   (* 180 radians, NOT 180 degrees. *)

−0.801153

N[Sin[180 Degree]]

0.

ArcTan[1]

$\dfrac{\pi}{4}$

Exp[Log[E]]

E

> **Notes**
>
> (1) All *Mathematica* built-in function names start with an upper-case letter and all use square brackets in syntax.
>
> (2) *Mathematica* uses radian measure for all angles.

Error Messages

If you make a mistake in your input, *Mathematica* will *beep* and/or *print an error message* when you evaluate it. In Version 3.0, the part of the expression that's in error will be underlined as well.

Mismatching parentheses is a common mistake. For example:

3*(4-5))+6

Syntax::bktmop :
  Expression "3 * (4 - 5))" has no opening "(".

3 * (4 - 5)) + 6

You can also make logical errors (not just typing errors) in your input. For example, dividing by zero in an expression doesn't make much sense:

5/0

Power::infy : Infinite expression $\dfrac{1}{0}$ encountered.

ComplexInfinity

(The expression "5/0" actually makes a lot of sense as a *complex* number.)

More Examples

Approximate Numbers and Exactness

Working with **approximate numeric values** is just like working with numbers on a hand-held calculator. These numbers sometimes lose precision as values are rounded off during computation.

■ **Example**. *Mathematica* makes a big distinction between the exact number π and a numerical approximation for it. For example, **Sin[517*Pi]** is an exact quantity with an exact answer:

Sin[517*Pi]

0

If you use a numerical approximation for 517π, you don't get an exact zero:

```
Sin[N[517*Pi]]
```

$$-6.03961 \times 10^{-14}$$

This is very close to zero (it is, after all, –0.0000000000000603961) and it's probably acceptable for the work you'll be doing. But it's not exact.

Useful Tips

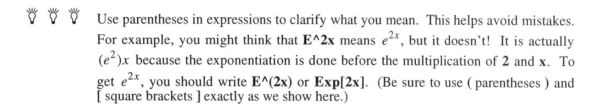

Use parentheses in expressions to clarify what you mean. This helps avoid mistakes. For example, you might think that **E^2x** means e^{2x}, but it doesn't! It is actually $(e^2)x$ because the exponentiation is done before the multiplication of **2** and **x**. To get e^{2x}, you should write **E^(2x)** or **Exp[2x]**. (Be sure to use (parentheses) and [square brackets] exactly as we show here.)

Avoid using the **%** symbol except for short sequences of one or two calculations.

A common mistake (for beginners) is to type **Ln[x]** for $\ln(x)$. You should use **Log[x]**.

Avoid using spaces for multiplication. Write ***** explicitly for multiplication.

Troubleshooting Q & A

Question... When I typed in **Sqrt[2]**, *Mathematica* just gave me the same thing back again. It didn't evaluate it. Why?

```
Sqrt[2]
```
```
Sqrt[2]
```

Answer... *Mathematica* always gives you an exact answer. When you write the square root of 2 as $\sqrt{2}$ in mathematics, you don't try to simplify it. Neither does *Mathematica*! However, if you want a decimal approximation for $\sqrt{2}$, use **N**.

```
Sqrt[2] // N
```
```
1.41421
```

Question... I entered a single command on more than one line, and *Mathematica* gave me more than one output! What happened?

Answer... Prior to version 3.0, *Mathematica* interpreted the **Return** key as the end of a command, as long as it made sense to do so. For example, try adding 1 and 2, but use two lines to do it:

```
1
+2

1
2
```

Mathematica took the first line to be **1**, which makes sense by itself. The second line also makes sense by itself. Thus, you get two results: 1 and 2.

To fix the problem, you either have to write **1+2** on the same line, or you can enter it this way:

```
1 +        (* 1+ is not a full expression, so the *)
2          (* 2 is considered its continuation. *)

3
```

This problem has been fixed in version 3.0.

Question...   After I input a built-in function or constant, *Mathematica* gave a warning message, "Possible spelling error." What went wrong?

Answer...   This means that you typed something that looks like (but is not the same as) the name of a built-in function or constant. Check your spelling. Most likely, you've not used upper- and lower-case correctly in spelling the name (e.g., **sin[pi]** instead of **Sin[Pi]**).

Question...   I got the wrong answer when I entered a complicated arithmetic expression. What should I check?

Answer...   Always use (*parentheses*) to keep your expressions manageable and readable. Sometimes you can get tripped up by not knowing the exact order in which expressions are evaluated. Parentheses make the order of evaluation clear.

For example, $(2^3)^4 = 4096$ and $2^{(3^4)}$ are very different numbers:

```
(2^3)^4

4096
```

```
2^(3^4)

2417851639229258349412352
```

But if you don't use parentheses, you might not know which one gets computed:

```
2^3^4          (* This evaluates to 2^(3^4). *)

2417851639229258349412352
```

CHAPTER 3
Variables and Functions

Variables

Immediate Assignment

With a calculator, you can store a value into memory and then recall it later. With a more advanced calculator, you can store different values under names such as A, B, C, and so on.

You can do even better in *Mathematica*. You can assign a name to any *Mathematica* expression or value and then recall it whenever you want. You do this using the equal sign "**=**", which is the symbol for **immediate assignment**. For example:

```
a = 3.4          (*We assign a to be the value 3.4.*)
3.4
```

Once you've made an assignment, you can recall its value or use it in an expression:

```
a
3.4

a + 2
5.4

a^2
11.56
```

How Expressions Are Evaluated

You may want to know how *Mathematica* keeps track of all the symbols and variables that you have defined.

Say we assign the name **myLunch** to the sum of **apple** and 3 times **banana**.

```
myLunch = apple + 3*banana
apple + 3 banana     (*Mathematica spits back the definition
                        because apple and banana have no
                        associated value.*)
```

Now suppose we give the value 2 to **apple** and then we ask that **myLunch** be reevaluated:

```
apple = 2            (*Now apple has the value 2.*)
2

myLunch
2 + 3 banana         (*When Mathematica reevaluates myLunch,
                        it has substituted the value 2 for apple.*)
```

If we now define the value of **banana** to be 3 and reevaluate **myLunch**:

banana = 3   (*Now **banana** has the value 3.*)

3

myLunch

11   (*When *Mathematica* reevaluates myLunch, it replaces
                       apple and banana by their respective values, and
                       simplifies the resulting expression. Bon appétite!!*)

Redefining and Clearing Symbols

Once a symbol name has been assigned a value or an expression, *Mathematica* retains the association until you end your session, redefine it, or explicitly clear it.

a = 3.4   (*We assign **a** to be 3.4.*)

3.4

a = 5   (*We reassign **a** to be 5.*)

5

a + 2   (*Mathematica* uses the new assigned value of **a**.*)

7

You tell *Mathematica* to forget about the assignment using the **Clear** command:

Clear[a]
a   (*We can see that **a** is no longer 5.*)

a

You can also **Clear** assignments for many names at the same time in one statement:

Clear[myLunch, apple, banana]

Rules for Names

Names you use can be made up of letters and numbers, subject to the following two rules:

- You can't use a name that begins with a number. For example, **2app** is not an acceptable name because *Mathematica* will interpret it as "2" times "**app.**"
- You can't choose names that conflict with *Mathematica*'s own names. For example, you can't name one of your own variables **Sin**.

All of the following are examples of legitimate names that you could use:

a, m, p1, A, area, Perimeter, Batman, good4you, classsOf2001

> **Note:** *Mathematica* distinguishes upper case and lower case characters. For example, the names **Batman**, **batman**, and **batMan** are different.

Substitution Command

You can substitute values into an expression without defining the variables explicitly. The substitution symbol "**/.**" is made using the slash and period symbols, with no space in between. It is used in the form:

expression /. { *list of substitutions using* **-> }**

For example, to substitute $x = 2$ and $y = 5$ into the expression $x^2 - 2xy$:

```
Clear[x, y]
x^2 - 2*x*y  /. { x -> 2, y -> 5}
-16
```

The arrow symbol "–>" is formed by entering the minus sign and greater than sign together, with no spaces in between. This is called a **substitution rule**. We'll use it a lot in later chapters.

One advantage of using the substitution command is that the value you substitute into a variable is temporary and is not remembered by *Mathematica*.

```
x              (*The value of x is unchanged after the previous substitution.*)

x
```

Finding the Larger Root of a Quadratic Equation

■ **Example**. We want to find the larger root of each of the following quadratic equations: $2x^2 + 5x - 6 = 0$ and $2x^2 + x - 3 = 0$.

The roots of a quadratic equation $ax^2 + bx + c = 0$, with $a \neq 0$, are found using the quadratic formula $\left(-b \pm \sqrt{b^2 - 4ac}\right)/(2a)$. The larger root, when $a > 0$, is thus:

```
Clear[a, b, c, largerroot]
largerroot = (-b + Sqrt[b^2 - 4*a*c] )/(2*a)
```
$$\frac{-b + \sqrt{b^2 - 4\,a\,c}}{2\,a}$$

Here are the larger roots of each of the two equations:

```
largerroot /. {a -> 2, b -> 5, c -> -6}
```
$$\frac{1}{4}\left(-5 + \sqrt{73}\,\right)$$

```
largerroot /. {a -> 2, b ->1, c ->-3}
1
```

Functions

Defining Functions

Mathematica has many built-in functions such as **N**, **Sqrt**, **Sin** and **Tan**. You can add your own functions as well.

For example, you can define the function $f(x) = x^2 + 5x$ in *Mathematica* with the expression below.

```
f[x_] := x^2 + 5*x
```

This syntax may seem awkward, but you should notice:

- An underscore character after the variable, **x_**, tells *Mathematica* that x is the variable of the function.

- The colon-equal sign := is a **delayed assignment command**. It behaves differently than "= " and *Mathematica* will not give any output for this statement.

Now we can do some evaluations:

```
f[7.1]
```
85.91

```
f[a]
```
$5\,a + a^2$

```
f[x+1]
```
$5\,(1+x) + (1+x)^2$

```
f[f[x+y]]
```
$5\left(5\,(x+y) + (x+y)^2\right) + \left(5\,(x+y) + (x+y)^2\right)^2$

Functions with More Than One Variable

Functions may have more than a single variable. A simple example is the computation of the average speed of an automobile.

If an automobile travels m miles in the span of t minutes, then its average speed in miles per hour is given by the expression $\frac{m}{(t/60)} = \frac{60m}{t}$. We then have a speed function " $f(m, t) = \frac{60m}{t}$ ":

```
Clear[speed]
speed[m_,t_] := 60m / t
```

If a distance of 45 miles is traveled in 30 minutes, the average speed will be 90 m.p.h.:

```
speed[45,30]
```
90

More Examples

Functions of Split Definition

Functions sometimes cannot be defined using a single formula. For example, the famous Heaviside function is defined by:

$$H(x) = \begin{cases} 1, & \text{if } x > 0 \\ 0, & \text{if } x \le 0 \end{cases}.$$

To define this function in *Mathematica*, we use the **Which** command as follows:

```
h[x_] := Which[ x > 0, 1 , x <= 0, 0]
```

(*The term "<=" in the command means ≤, less than or equal to.*)

To use **Which** to define a function of split definition having two branches, write:

```
Which[ condition1 , result1 , condition2 , result2 ]
```

This means that if *condition1* is satisfied, then *result1* will be used; otherwise, *Mathematica* will use *result2* if *condition2* is satisfied. The following table shows some operators you'll use to check conditions.

Operator	Meaning		Operator	Meaning
==	equal to		!=	not equal to
>	greater than		<	less than
>=	greater than or equal to		<=	less than or equal to
&&	and		||	or

> **Note:** The symbols ==, !=, >=, <=, && and || (two vertical lines) are formed by typing two characters without any space in between.

The syntax for the **Which** command can also be extended. For example, the following function *f* has three branches and can be defined as you see below:

$$f(x) = \begin{cases} 1 - x, & \text{if } 1 < x < 3 \\ x^2, & \text{if } 0 \le x \le 1 \\ x + 2, & \text{if } x < 0 \text{ or } x \ge 3 \end{cases}$$

```
f[x_] := Which[ 1 < x < 3, 1-x, 0 <= x <= 1, x^2,
                      x <0 || x>=3, x+2]
```

To see this definition more clearly, we recommend that you write the conditions one on each line and line them up carefully:

```
f[x_] := Which[   1 < x < 3,        1-x,
                  0 <= x <= 1,      x^2,
                  x <0 || x>=3,     x+2]
```

Useful Tips

💡 💡 💡 💡 Never assign values to any of the names **x**, **y**, **z**, or **t**. Otherwise, *Mathematica* will confuse them with the variables **x**, **y**, **z**, or **t** that you typically use when defining functions.

💡 💡 💡 💡 All of *Mathematica*'s built-in names begin with a capital letter. Start your names with a lower-case letter so you'll be able to distinguish yours from *Mathematica*'s.

💡 💡 💡 💡 You should always use **Clear** before defining functions. **Clear** helps to avoid potential conflicts between variable names and function definitions.

💡💡 You can use the command **Clear["@"]** to clear all names you created that are made up only of lower-case characters and numbers.

Troubleshooting Q & A

Question... When I defined a new variable, I got the warning message "`Possible spelling error.`" What did I do wrong?

Answer... When you define a name that looks a lot like another name you used before (e.g., **potatos** and **potatoes**), *Mathematica* warns you about it. This is meant to be a helpful reminder, letting you know that you might have mistyped the name. You didn't do anything wrong and *Mathematica* still accepted your definition.

Mathematica won't complain about using the names **potato1**, **potato2** and **potato3**.

Question... I tried to define a function, but I couldn't get it to work. What should I check?

Answer... Two things usually bother function definitions.

- First, check for the proper syntax. Use an underscore *immediately* following the variable, use a colon-equal definition, and don't use any underscores on the right side of the colon-equal symbol.
- Second, your function or variable name may conflict with something else you used earlier in your *Mathematica* session. Type **Clear[f, x]** before you define your function.

Question... Why didn't my definition of a function **f** replace my earlier definition?

Answer... This happened in *Mathematica* 2.2 when you used different variables in defining a function. For example:

```
f[x_] := x^2
f[x_] := x + 2        (*This will replace the earlier definition x^2.*)
f[t_] := 3t + 5       (*This will not replace the definition x+2,
                          because the variables x and t are different.*)
f[t]
t+2                   (*Mathematica is still using the old definition.*)
```

You can avoid all these mistakes by typing **Clear[f]** prior to defining the function *f*. Luckily, this trouble is fixed in version 3.0.

Question... I'm still not sure about when I should use an equal sign =, or when I should use a colon-equal := definition? What's the easiest way to remember?

Answer... The symbol = is **immediate assignment**, while **:=** is **delayed assignment**. The difference between them is about when the expression on the right side of the assignment gets evaluated. Our rule of thumb is:

- Use immediate assignment "**=**" when you give a name to an expression or result.
- Use delayed assignment "**:=**" when you define a function.

CHAPTER 4
Computer Algebra

Working with Polynomials and Powers

The Expand and Factor Commands

The **Expand** command does exactly what its name says it does:

> `Expand[(x-2)(x-3)(x+1)^2]`
>
> $6 + 7x - 3x^2 - 3x^3 + x^4$

The **Factor** command is basically the reverse of the **Expand** command:

> `Factor[6+7*x-3*x^2-3*x^3+x^4]`
>
> $(-3 + x)(-2 + x)(1 + x)^2$

Here are a few more examples:

Example	Comment
`Factor[x^2-3]` $-3 + x^2$	Although $x^2 - 3 = (x + \sqrt{3})(x - \sqrt{3})$, **Factor** will not give radicals in its answer.
`Expand[(x-1.54)(3.2x-2.9)]` $4.466 - 7.828x + 3.2x^2$ `Factor[%]` $3.2(-1.54 + x)(-0.90625 + x)$	**Factor** does a nice job even when you use numerical coefficients.
`Factor[(5-3I)+(-4+I)x+(1-I)x^2]` $((1 + I) - Ix)((1 - 4I) + (1 + I)x)$	**Factor** works even when the coefficients are complex.
`Factor[x^2+1]` $1 + x^2$	**Factor** won't use complex numbers unless at least one of the coefficients is a complex number.
`Expand[(x-y+z)^3]` $x^3 - 3x^2y + 3xy^2 - y^3 + 3x^2z - 6xyz + 3y^2z + 3xz^2 - 3yz^2 + z^3$ `Factor[%]` $(x - y + z)^3$	The **Expand** and **Factor** commands also work for polynomials with more than one variable.

The Simplify Command

The **Simplify** command tries to produce an expression that is the shortest to write out. Most often, this will be the same as what you think of as "simplest."

For example, $x^2 - 2x + 1 = (x - 1)^2$ is factored when you **Simplify** it:

> `Simplify[x^2-2x+1]`

$$(-1 + x)^2$$

Even though $x^3 + 2x^2 - 2x - 1$ factors as $(x-1)(x^2 + 3x + 1)$, *Mathematica* thinks that writing it as a polynomial is simpler than factoring it (notice we use the postfix syntax **// Simplify** here):

```
x^3+2x^2-2x-1 // Simplify
```

$$-1 - 2x + 2x^2 + x^3$$

The PowerExpand Command

In some cases, you can use the **PowerExpand** command to "handle" algebra involving exponents. For example, $\sqrt{x^2}$ cannot be simplified by either the **Simplify** or **Expand** commands.

```
Simplify[Sqrt[x^2]]
```

$$\sqrt{x^2}$$

```
Expand[Sqrt[x^2]]
```

$$\sqrt{x^2}$$

(Note that $\sqrt{x^2} = x$ only when $x \geq 0$.)

However, the **PowerExpand** command will treat $\sqrt{x^2}$ as $(x^2)^{1/2}$ and then rewrite it to be $x^{2(1/2)} = x^1$.

```
PowerExpand[Sqrt[x^2]]
```

```
x
```

Similarly:

```
Simplify[(x^6)^(1/3)]
```

$$\left(x^6\right)^{1/3}$$

```
PowerExpand[(x^6)^(1/3)]
```

$$x^2$$

Working with Rational Functions

Together and Apart

A rational function is an expression of the form $\frac{a\ polynomial}{another\ polynomial}$. The following are three common algebraic operations involving rational functions.

(i)   Combining terms over a common denominator is done with the **Together** command. For example, to combine $\frac{2}{3x+1} + \frac{5x}{x+2}$:

```
Together[ 2/(3x+1) + (5x)/(x+2)]
```

$$\frac{4 + 7\,x + 15\,x^2}{(2 + x)\,(1 + 3\,x)}$$

(ii) Splitting up rational functions into partial fractions is done with the **Apart** command. For example, to split up $\dfrac{11x^2 - 17x}{(x-1)^2(2x+1)}$:

Apart[(11x^2-17x)/((x-1)^2*(2x+1))]

$$-\frac{2}{(-1+x)^2} + \frac{3}{-1+x} + \frac{5}{1+2\,x}$$

(iii) The **Apart** command also does long division. For example, you can compute the quotient $(x^5 - 2x^2 + 6x + 1) \div (x^2 + x + 1)$ with:

Apart[(x^5-2*x^2+6x+1)/(x^2+x+1)]

$$-1 - x^2 + x^3 + \frac{2 + 7\,x}{1 + x + x^2}$$

Working with Trig and Hyperbolic Functions

Basic Trig and Hyperbolic Identities

The **Simplify** command recognizes most basic identities involving the trigonometric and hyperbolic functions:

Simplify[Sin[x]^2+Cos[x]^2]

1

Simplify[2Sin[x]Cos[x]]

Sin[2x]

Simplify[Cosh[x]^2-Sinh[x]^2]

1

Simplify[Sinh[x]^2+Cosh[x]^2]

Cosh[2 x]

Trig Option in the Factor and Expand Commands

If you want to use **Factor** or **Expand** effectively with trig or hyperbolic functions, you need to add the option **Trig –> True**. (The arrow "–>" is formed by typing the minus sign "–" and greater than sign ">" together, with no space in between.)

Factor[Sin[2x], Trig->True]

2 Cos[x] Sin[x]

Expand[Sin[2x]Cos[3x], Trig->True]

$2\,\text{Cos}[x]^4\,\text{Sin}[x] - 6\,\text{Cos}[x]^2\,\text{Sin}[x]^3$

> **Note:** Version 3.0 of *Mathematica* has new commands **TrigExpand**, **TrigFactor** and **TrigReduce** which give more control in working with trigonometric functions. They automatically assume **Trig –> True**.

The FullSimplify Command

In version 3.0 of *Mathematica*, there's a new, "full-strength" version of **Simplify** named **FullSimplify**. This command tries a wider variety of expression simplification possibilities than **Simplify** (but it may take a long time with complicated expressions). It's often useful when you're working with transcendental functions.

For example, consider the identity $\dfrac{e^{\tanh^{-1} x} - e^{-\tanh^{-1} x}}{e^{\tanh^{-1} x} + e^{-\tanh^{-1} x}} = x$:

(Exp[y]-Exp[-y])/(Exp[y]+Exp[-y]) /. y->ArcTanh[x]

$$\frac{-E^{-\text{ArcTanh}[x]} + E^{\text{ArcTanh}[x]}}{E^{-\text{ArcTanh}[x]} + E^{\text{ArcTanh}[x]}}$$

Simplify only makes a minor rearrangement of the expression but can't really "simplify." **FullSimplify** gets the job done!

Simplify[%]

$$\frac{-1 + E^{2\,\text{ArcTanh}[x]}}{1 + E^{2\,\text{ArcTanh}[x]}}$$

FullSimplify[%]

x

Useful Tips

 You have to be careful when you use **PowerExpand**. Whenever you use it, you will usually be assuming that all the quantities you're working with are nonnegative real numbers.

Troubleshooting Q & A

Question...   When I used **Simplify**, **Expand**, or **Factor** on a polynomial, I got a number. What went wrong?

Answer...   First of all, after expansion and simplification, it is possible that all the terms in your polynomial canceled out and left you with a constant.

If this is not the case, check whether you assigned a value to a variable at some earlier time. For example, you may have assigned

x = 5

earlier; then after awhile you typed:

```
Expand[(x-3)^2*(4*x+5)]
100
```

You asked *Mathematica* to expand (5-3)^2*(4*5+5) = 100!! To correct this, type:

```
Clear[x]
```

and reenter the **Expand** command.

Question...   I tried **Simplify**, **Factor**, and several other commands on an expression, and could not get the form I was expecting. What should I do?

Answer...   If you're working with trigonometric or hyperbolic functions, you might try using the option **Trig –> True**. If you have version 3.0 of *Mathematica*, try **FullSimplify**. Otherwise, the answer may be that the software just can't give you what want using algebra alone. You must look for a different approach.

For example, *Mathematica* can't simplify $\tanh^{-1}\!\left((e^x - e^{-x})/(e^x + e^{-x})\right) = x$. But if you graph the function $\tanh^{-1}\!\left((e^x - e^{-x})/(e^x + e^{-x})\right)$, you see that it looks like $y = x$. (Chapter 8 does graphing.)

You can also compute that $\tanh^{-1}\!\left((e^x - e^{-x})/(e^x + e^{-x})\right)$ has derivative 1, so that's almost enough to establish the identity. (Chapter 11 will show you how to use derivatives.)

Question...   I tried to use the **Trig** option with one of the **Factor** or **Expand** commands, but I got one of the error messages "Symbol Trig is Protected" or "Trig ... cannot be followed by ... True." What did I do wrong?

Answer...   Check that you typed the option **Trig –> True** correctly. A common mistake is to type **Trig = True** (with an equal sign).

Also, when you form the arrow "**–>**", make sure that there is no space between the minus sign "**–**" and greater-than sign "**>**".

CHAPTER 5
Working with Equations

Equations and Their Solutions

The Solve Command for an Equation

Mathematica's **Solve** command will solve an equation for an unknown variable. You use it in the form:

> **Solve[** *an equation* **,** *variable to solve for* **]**

For example,

> **Solve[2x+5 == 9, x]**
>
> {{x -> 2}}

Notice that

- Equations in *Mathematica* are written using the double-equal sign "==". (Single = means assignment as we mentioned in Chapter 3.)

- The output of the answer involves "->", which is used with substitution rules (also described in Chapter 3).

You can check that $x = 2$ is the correct solution to the above equation, using substitution syntax:

> **2x+5 == 9 /. { x -> 2 }**
>
> True

This means that after the substitution $x = 2$, it is "True" that the left-hand side of the equation equals the right-hand side.

Here are a few more examples involving **Solve**:

Equation	*To Solve It in* Mathematica	*Comment*
Solve $x^2 - 3x + 1 = 0$ for x	**Solve[x^2-3x+1 == 0, x]** $$\left\{\left\{x \to \frac{1}{2}\left(3-\sqrt{5}\right)\right\}, \left\{x \to \frac{1}{2}\left(3+\sqrt{5}\right)\right\}\right\}$$ **N[%]** $\{\{x \to 0.381966\}, \{x \to 2.61803\}\}$	Equations can have more than one solution. We can see a numerical answer with the **N** command.
Solve $x^3 + x^2 = -3x$ for x	**Solve[x^3+x^2==-3x,x]** $$\left\{\{x \to 0\}, \left\{x \to \frac{1}{2}\left(-1 - I\sqrt{11}\right)\right\},\right.$$ $$\left.\left\{x \to \frac{1}{2}\left(-1 + I\sqrt{11}\right)\right\}\right\}$$	Here two of the solutions are complex. **I** stands for $\sqrt{-1}$.

Solve $y^2 - ay = 2a$ for y.	`Solve[y^2 - a*y == 2a, y]` $\left\{\left\{y \rightarrow \frac{1}{2}\left(a - \sqrt{a}\sqrt{8+a}\right)\right\},\right.$ $\left.\left\{y \rightarrow \frac{1}{2}\left(a + \sqrt{a}\sqrt{8+a}\right)\right\}\right\}$	If the equation involves other variables, *Mathematica* will treat them as constants.
Solve $x + \sin x = \cos x$ for x	`Solve[x+Sin[x] == Cos[x], x]` `Solve[x + Sin[x] == Cos[x], x]`	The input is returned unevaluated when *Mathematica* cannot solve an equation.

> **Note:** The **Solve** command works very well for equations involving polynomials. However, it doesn't have much success with trigonometric, exponential, logarithmic, or hyperbolic functions.

The Solve Command for a System of Equations

Solve can also be used to solve a system of equations. For two equations in two unknown variables, you use this form of the command:

> `Solve[{ equation1, equation2 }, { variable1, variable2 }]`

■ **Example.** To solve the equations $3x + 8y = 5$ and $5x + 2y = 7$ in the variables x and y, use:

> `Solve[ {3x+8y == 5, 5x+2y == 7} , {x,y} ]`
> $\left\{\left\{x \rightarrow \frac{23}{17}, y \rightarrow \frac{2}{17}\right\}\right\}$

To solve the equations $3xy - y^2 = -4$ and $2x + y = 3$:

> `Solve[ {3x*y - y^2 == -4, 2x + y == 3}, {x,y} ]`
> $\left\{\left\{x \rightarrow \frac{1}{20}\left(21 - \sqrt{241}\right), y \rightarrow \frac{1}{10}\left(9 + \sqrt{241}\right)\right\},\right.$
> $\left.\left\{x \rightarrow \frac{1}{20}\left(21 + \sqrt{241}\right), y \rightarrow \frac{1}{10}\left(9 - \sqrt{241}\right)\right\}\right\}$

If you try:

> `Solve[{x+y == 0, x+y==1},{x,y}]`
> `{}`

This means that there is no solution to this system of equations.

We can use a similar syntax to solve systems of equations involving three or more variables. For example:

> `Solve[ {x+2*y-z==1, x-y+z^2==2, y+x==2z}, {x,y,z}]`
> $\left\{\left\{x \rightarrow -7 - 6\sqrt{2}, y \rightarrow 3 + 2\sqrt{2}, z \rightarrow 2\left(-1 - \sqrt{2}\right)\right\},\right.$
> $\left.\left\{x \rightarrow -7 + 6\sqrt{2}, y \rightarrow 3 - 2\sqrt{2}, z \rightarrow 2\left(-1 + \sqrt{2}\right)\right\}\right\}$

Equations and Numerical Solutions

The NSolve Command for an Equation

The **NSolve** command uses efficient numerical techniques to approximate roots of polynomials and a few other simple functions.   It has the same syntax as **Solve**.

NSolve[*an equation* , *variable to solve for*]

```
NSolve[x^5-2x^3-1==0, x]
```

$\{\{x \to -1.17872\}, \{x \to -1.\}, \{x \to 0.332924 - 0.670769 \text{ I}\},$
$\{x \to 0.332924 + 0.670769 \text{ I}\}, \{x \to 1.51288\}\}$

The NSolve Command for Systems of Equations

NSolve can also do systems of equations, just like **Solve**. You use the form:

NSolve[{ *equation1*, *equation2* }, { *variable1*, *variable2* }]

For example,

```
NSolve[ {3x*y - y^2 == 4, 2x^2 + y == 9}, {x,y} ]
```

$\{\{x \to -2.94619, y \to -8.36012\}, \{x \to -2.19947, y \to -0.675324\},$
$\{x \to 1.61445, y \to 3.78713\}, \{x \to 2.03122, y \to 0.748316\}\}$

If you tried **Solve** with the system of equations above, it would take about 100 times longer than **NSolve** takes and several pages to print out the symbolic solutions!

FindRoot and Numerical Solutions

The FindRoot Command

NSolve doesn't work at all with exponential, logarithmic, trigonometric, or hyperbolic functions. To solve these types of equations numerically, you should use **FindRoot**. You use it in the form:

FindRoot[*equation* , { *variable*, *first estimate of a solution* }]

FindRoot will try to find a solution that usually is close to your first estimate.

■ **Example.** Let's find a solution of $\tan(x) = 8 - 17x^2$ using **FindRoot**. If we know that there is a solution near $x = 0.6$, then we can type:

```
FindRoot[ Tan[x] == 8-17x^2, {x, 0.6} ]
```

$\{x \to 0.652415\}$

We can see that there's a solution near $x = 0.6$ by looking at the intersections of the graphs of $\tan(x)$ and $8 - 17x^2$. You'll learn how to do this in Chapter 8, but for now, here's the picture on the right.

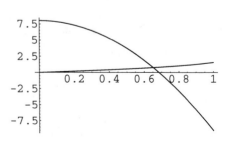

FindRoot and Systems of Equations

To solve a system of two equations in the two unknowns x and y, using $x = x_0$ and $y = y_0$ as initial estimates of a solution, you write:

```
FindRoot[{ equation1, equation2 }, { x , x₀ }, { y , y₀ }]
```

■ **Example.** The equations $y^2 - x^3 = 5$ and $y = x - 3\cos x + 4$ have a simultaneous solution that's close to $x = 1$ and $y = 2$. We can pinpoint it with:

```
FindRoot[ {y^2-x^3==5,y==x-3Cos[x]+4}, {x,1}, {y,2} ]
```

$\{x \to 0.663687, y \to 2.30051\}$

More Examples

Extracting Solutions from the Results of Solve

Sometimes you will want to work with the answers you get from **Solve** and **NSolve** without having to retype them. This example will show you the three main steps to do this.

Consider the equation $x^2 - 3x + 1 = 0$. *Mathematica* gives you two solutions:

```
Solve[ x^2-3x+1 == 0, x ]
```

$$\left\{ \left\{ x \to \frac{1}{2} \left(3 - \sqrt{5} \right) \right\}, \left\{ x \to \frac{1}{2} \left(3 + \sqrt{5} \right) \right\} \right\}$$

Notice that the output above has two substitution rules that describe the answers.

• Step 1. Give the output of the **Solve** command a name:

```
ans = Solve[ x^2-3x+1 == 0, x ]
```

$$\left\{ \left\{ x \to \frac{1}{2} \left(3 - \sqrt{5} \right) \right\}, \left\{ x \to \frac{1}{2} \left(3 + \sqrt{5} \right) \right\} \right\}$$

• Step 2. Identify each of the answers with **ans[[1]]** and **ans[[2]]**.

```
ans[[1]]
```

$$\left\{ x \to \frac{1}{2} \left(3 - \sqrt{5} \right) \right\}$$

```
ans[[2]]
```

$$\left\{ x \to \frac{1}{2} \left(3 + \sqrt{5} \right) \right\}$$

Notice that each of the answers is a substitution rule.

• Step 3. Extract the values from the substitution rules. For example, here's how to get the second value:

```
root2 = x /. ans[[2]]
```

$$\frac{1}{2} \left(3 + \sqrt{5} \right)$$

In the command above, *Mathematica* evaluated the "dummy variable" **x** using the substitution rule $\{x-> \frac{1}{2}(3+\sqrt{5})\}$. This assigned $\frac{1}{2}(3+\sqrt{5})$ to **root2**.

Let's check that **root2** is really a solution to $x^2 - 3x + 1 = 0$:

```
(root2)^2 - 3(root2) + 1 // Simplify
```
0

Useful Tips

♀ ♀ ♀   **NSolve** works much more quickly using numeric methods than **Solve** does algebraically. In fact, **Solve** cannot solve many equations symbolically! Use **NSolve** whenever you can.

♀ ♀ ♀   **Clear** all the variables that you want to solve for, before you enter the equation(s) in a **Solve** or **NSolve** command.

♀ ♀   Both **NSolve** and **FindRoot** work quickly, but you'll usually find that **NSolve** is more convenient because you don't have to come up with a first approximation. **NSolve** is also more accurate than **FindRoot**. But **NSolve** doesn't work all the time.

♀   The following table can help you remember the difference between various types of brackets. But beware – they are not interchangeable!

Syntax Element	Purpose	Example
(*parentheses*)	Grouping of terms	`(3*x+1)^2+(y+2)`
[*square brackets*]	Function arguments	`Sin[x]`
{ *curly braces* }	Lists	`{x+1, x}`
[[*double square brackets*]]	Specify by position	`ans[[1]]`

Troubleshooting Q & A

Question...   I tried to **Solve** some simple equations, but I kept getting errors. What should I look for to fix my input?

Answer...   Usually, one of two common errors are made when using the **Solve** command.

- The first thing to check is syntax. Make sure you are using two equal signs, with no spaces in-between, to enter each equation. If you are using more than one equation, make sure all the curly braces and commas are located in the right places.

- Second, make sure the variable or variables you are trying to solve for have no assigned values. **Clear** them before using **Solve**.

Question... I got the error message "`... is not a well-formed equation`" when I used **Solve** or **NSolve**. What should I check?

Answer... Make sure you entered the entire equation, including the double equal sign and both sides of the equation.

Question... I used **Solve** but could not understand *Mathematica*'s output. It had one of these symbols:

- `ToRules`
- `Root`
- `#1.`

What do these mean?

Answer... This means that *Mathematica* cannot show you the solution explicitly. (The exact answers may be too complicated to display.) If the solution does not involve symbolic constants, you can use **N[%]** to see the numerical value of the answers.

Question... I got the output "`{ }`" from **Solve** or **NSolve**. What does this mean?

Answer... This means that there is no solution to your equation or system of equations. A single equation might reduce to an absurdity (e.g., $x = x + 1$ reduces to $0 = 1$). A system of equations may be inconsistent (e.g., $x + y = 1$ and $x + y = 2$).

Question... I used both **Solve** and **NSolve** with an equation, but my input came back unevaluated with an error message or two. What should I do?

Answer... **Solve** and **NSolve** work best on polynomials and rational functions, and don't work much otherwise. For example, to solve $\cos(x) = x$, *Mathematica* can only "take the inverse cosine of both sides" to get $x = \cos^{-1}(x)$. As you see, this equation isn't much better, and you will get an error message.

To solve this equation, use **FindRoot**. For example, there's a root of $\cos(x) = x$ near $x = 1$:

```
FindRoot[Cos[x]==x,{x,1}]
{x -> 0.739085}
```

Question... I got the warning that "`... Equations may not give solutions for all "solve" variables.`" What does this mean?

Answer... This usually happens when there are not enough equations to solve for all variables uniquely. Some of the variables will be treated as parameters, and solutions will be expressed in terms of them.

Question... **FindRoot** gave me a ridiculous answer to a simple equation. What happened?

Answer... **FindRoot** uses an iterative method to solve the equation. The procedure stops when the equation is satisfied to a reasonable numerical precision. A typical example is:

```
FindRoot[ x^10 == 0, {x, 0.2} ]
```

```
{x -> 0.18}
```

Obviously, $x = 0$ is the only solution to this equation. But $(0.18)^{10} \approx 0$ to seven digits of accuracy, so *Mathematica* takes $x = 0.18$ to be a reasonable solution.

Question...   When I used **FindRoot**, I got the message that "`...  Newton's  method failed to converge to the prescribed accuracy ...`", and then I got an answer anyway. Can I trust the answer?

Answer...   No. You should retry the **FindRoot** command with a better first approximation for the root. The one you used originally was probably too far away from an actual root of the equation.

CHAPTER 6
Lists and Tables

Lists

What Is a List?   A **list** in *Mathematica* is an expression whose elements are separated by commas and enclosed in curly braces. For example, each of the following is a list.

> `{2, 5, 7, 10, -3, -25}`   (*Each element is a number.*)

> `{"good", "bad", "ugly"}`   (*Each element is a name or string.*)

> `{{1,3}, {2,1}, {5,6}}`   (*Each element is a pair of numbers.*)

Lists are important structures in *Mathematica*. Many of *Mathematica*'s inputs and outputs are expressed using lists. For example, both the input and output syntax used in solving a system of equations involves the use of lists:

> `Solve[ {x+y==2, x^2+y == 2}, {x,y}]`
> `{{y -> 1, x -> 1}, {y -> 2, x -> 0}}`

Basic List Operations   It makes sense to add, subtract, multiply, and divide lists in *Mathematica*. The following table shows some examples of operations you can perform with lists:

Mathematica *Command*	Explanation
`{1, 3, 5, 7, 22} + 6` `{7, 9, 11, 13, 28}`	Adds 6 to each element in the list.
`{2, 5, 1, 2, 1,-2} * 3` `{6, 15, 3, 6, 3, -6}`	Multiplies each element in the list by 3.
`{1, 2, 4, -2, 6} ^ 3` `{1, 8, 64, -8, 216}`	Raises each element in the list to the power 3.
`{1, 2, 3} + {-3, 0, 5}` `{-2, 2, 8}`	If two lists have the same number of elements, we can add them element by element.
`{1, 3, 5} * {2, -5, 2}` `{2, -15, 10}`	If two lists have the same number of elements, we can multiply them element by element.
`{3, -2, 1, 5, 6}[[4]]` `5`	We can use [[*square brackets*]] to pick out a specific element in the list. In this example, we pick out the fourth element.

Lists and Substitution   ■ **Example**. We're given the functions $f(x) = x^2 - 5x + 1$ and $g(x) = 2\sin(x)$:

> `f[x_] := x^2 - 5x + 1`
> `g[x_] := 2Sin[x]`

Suppose we want to find the values of both f and g at $x = 1.0$, 1.5, 3.25, 5.1 and 6.5. We can do this very quickly with the help of a list:

```
list1 = {1.0, 1.5, 3.25, 5.1, 6.5}
```

```
{1, 1.5, 3.25, 5.1, 6.5}
```

    f[x] /. x -> list1   (*We substitute $x = 1.0$, 1.5 etc. into **f[x]**.*)

```
{-3., -4.25, -4.6875, 1.51, 10.75}
```

    g[x] /. x-> list1   (*We substitute $x = 1.0$, 1.5 etc. into **g[x]**.*)

```
{1.68294,1.99499,-0.21639,-1.85163,0.43024}
```

You'll see more commands and techniques for working with lists in Chapter 24.

Tables

The Table Command

If the elements of a list can be defined by a mathematical formula, we can create the list using the **Table** command. For example, to make a list where each element is of the form n^2, for each integer $1 \le n \le 20$, we will type:

```
Table[ n^2, {n, 1, 20}]
```

```
{1,4,9,16,25,36,49,64,81,100,121,144,169,196,225,
 256,289,324,361,400}
```

In general, the **Table** command is used in the form:

> **Table[** *expression*, **{n,** n_0, n_1**}]**

or, if we want **n** to increase from n_0 to n_1 in steps of m, we will use

> **Table[** *expression*, **{n,** n_0, n_1, m**}]**

Here are a few examples:

Mathematica *Command*	*Remark*
Table[n^2,{n,3,10}] {9,16,25,36,49,64,81,100}	The variable n runs from 3 to 10.
Table[n^2,{n,3,10,2}] {9,25,49,81}	The variable n runs from 3 to 10 in steps of 2.
Table[n^2,{n,-5,5,2.5}] {25,6.25,0.,6.25,25.}	The variable n runs from −5 to 5 in steps of 2.5.
Table[{Cos[n],n/(n+1)},{n,1,5}] $\{\{Cos[1], \frac{1}{2}\}, \{Cos[2], \frac{2}{3}\}, \{Cos[3], \frac{3}{4}\},$ $\{Cos[4], \frac{4}{5}\}, \{Cos[5], \frac{5}{6}\}\}$	The expression in the **Table** command can be a list itself. In this case, we form a table of coordinate pairs.
Table[x^n,{n,0,7}] $\{1, x, x^2, x^3, x^4, x^5, x^6, x^7\}$	Each element can be a symbolic expression.

Plotting Points

ListPlot Command

If each element of a list is a number, we can use **ListPlot** to see the numbers:

```
ListPlot[{1, 4, -2, 1, 4, -3, 2, 7}]
```

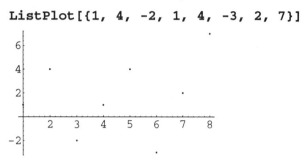

PlotStyle Options

In the preceding graphic, the points are too small to stand out. You can make them larger with the option **PlotStyle –> PointSize[0.03]**. (This means the size of each point is about 3% of the width of the graphic.)

```
ListPlot[{1, 4, -2, 1, 4, -3, 2, 7},
        PlotStyle -> PointSize[0.03]]
```

We can also connect the points using **PlotJoined –> True**.

■ **Example**. The following picture shows the Dow Jones industrial averages on the 12 trading days between November 3, 1997 and November 18, 1997:

```
djAvgs = {7674.39, 7689.13, 7692.57, 7683.24,
          7581.32, 7552.59, 7558.73, 7401.32,
          7487.76, 7572.48, 7698.22, 7650.82}
ListPlot[djAvgs , PlotJoined -> True,
        AxesLabel -> {"days", "Dow Jones Averages"}]
```

Note that we added the option **PlotJoined –> True** to connect the points and used the option **AxesLabel** to label the axes.

Plotting Pairs of Coordinates

If you have a list of *pairs* of numbers, for example:

```
list1 = {{0.5, 2},{3, 5},{-2, 1.5},{-1, 4},{2, -3.1}}
```

then this is recognized by **ListPlot** as a list of *x*- and *y*-coordinates of points in the plane. Hence the points can be plotted with:

```
ListPlot[list1, PlotStyle -> PointSize[0.03]]
```

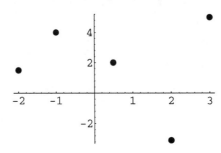

More Examples

An Experiment in Factoring

Let's look at the factorization of $x^n + 1$. With the help of the **Table** command, we can see the factorization for several values of *n* very quickly, say for $n = 2, 3, \ldots, 10$.

```
Table[ Factor[x^n+1], {n, 2, 10} ]
```

$$\{1 + x^2, \ (1 + x)\left(1 - x + x^2\right), \ 1 + x^4,$$
$$(1 + x)\left(1 - x + x^2 - x^3 + x^4\right), \ \left(1 + x^2\right)\left(1 - x^2 + x^4\right),$$
$$(1 + x)\left(1 - x + x^2 - x^3 + x^4 - x^5 + x^6\right),$$
$$1 + x^8, \ (1 + x)\left(1 - x + x^2\right)\left(1 - x^3 + x^6\right),$$
$$\left(1 + x^2\right)\left(1 - x^2 + x^4 - x^6 + x^8\right)\}$$

Do you notice that all the coefficients in all of the factors above are either 1 or –1? Try repeating the command with **{ n, 2, 100 }**. You'll see the same pattern!

You may probably conclude that every coefficient in every factor of $x^n + 1$ will be ±1, no matter what *n* is. But before you celebrate your latest discovery, check the factorization of $x^{105} + 1$. Surprise!

Drawing a Nonagon

■ **Example**. A nonagon is a regular polygon of nine sides (i.e., all nine sides have the same length, and all nine angles are equal). It can be formed by joining together the points with coordinates $(\cos(2\pi n/9), \sin(2\pi n/9))$, where $n = 1, 2, \ldots, 9$.

We can create a list of these points using the **Table** command and then use **ListPlot** with **PlotJoined –> True** to join the points.

```
ListPlot[Table[{Cos[n*2*Pi/9],Sin[n*2*Pi/9]},
    {n,0,9}], PlotJoined ->True,
    AspectRatio -> Automatic]
```

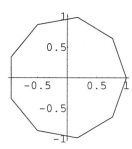

(The option **AspectRatio –> Automatic** is used to make sure that both *x*- and *y*-axes have the same unit scale. We will explain this in detail in Chapter 8.)

Troubleshooting Q & A

Question... When I used **Table**, *Mathematica* gave me the output { }. What does that mean?

Answer... This means that the list you asked for is empty, because the index range you specified does not make sense. Make sure that $n_0 \leq n_1$ in the command

> **Table[** *expression,* **{n,** n_0**,** n_1**}]**

Question... I got an error message when I used **ListPlot**. What should I check?

Answer... The three most likely things to check are:

- Make sure that all the curly braces and commas in your list of points are in the right places.
- Check that you used the correct syntax for **ListPlot**. If you added options, make sure that they are spelled correctly.
- **ListPlot** will only work for lists that are made up of numbers, or pairs of numbers. You cannot mix the two types of lists. No elements of your list can be symbolic – they must evaluate as numerical values.

Question... I used **ListPlot** with both the **PlotJoined–>True** and **PlotStyle–>PointSize[0.03]** options, but the points don't show up thick in the output. What happened?

Answer... **ListPlot** cannot handle both options at the same time (although we don't know why!). But you can get both effects on the same output if you follow this sequence:

> **pict1 = ListPlot[** *your list,* **PlotJoined->True]**
> **pict2 = ListPlot[** *your list,* **PlotStyle->PointSize[0.03]]**
> **Show[pict1,pict2]**

We'll explain more about the **Show** command in Chapter 8.

CHAPTER 7

Getting Help and Loading Packages

Text-based Help

Help for Specific *Mathematica* Commands

No matter which version of *Mathematica* you use, you can always get some help on how to use a command by typing:

> **?** *command*

For example, if you forget how to use the **FindRoot** command, you can type:

> **?FindRoot**
>
> ```
> FindRoot[lhs == rhs, {x, x0}] searches for a numerical solution
>     to the equation lhs == rhs, starting with x == x0.
> ```

You may not be comfortable with this message. (The terms `lhs` and `rhs` are short-hand for the *left-hand side* and the *right-hand side* of the equation.) But you can at least pick out that the general form of the command is:

> **FindRoot[** *equation* **, {** *variable* **,** *first estimate of the solution* **}]**

If you type **??** with a command name, you get more information. For example:

> **??Factor**
>
> ```
> Factor[poly] factors a polynomial over the integers.
>     Factor[poly, Modulus->p] factors a polynomial modulo a
>     prime p.
> ```
>
> ```
> Attributes[Factor] = {Protected}
> ```
>
> ```
> Options[Factor] = {GaussianIntegers -> False,  Modulus -> 0,
>     Trig -> False}
> ```

Mathematica shows you how to use the command and lists its options. To find out more about the option `Modulus`, for example, you can type:

> **?Modulus**
>
> ```
> Modulus is an option that can be given to Factor and other
>     mathematical operations. Modulus -> n specifies that numbers
>     should be treated modulo n. Modulus -> 0 specifies that the
>     full ring of integers should be used.
> ```

Looking Up a Command

Sometimes, you won't remember the exact name of a command that you want to use. Say, you remember that there's some type of **Root** command that you've used before, but you just can't recall its full name. You can ask *Mathematica*:

?*Root*

FindRoot   NRoots   Roots

The asterisk "*****" is a **wild-card character** in the command above. The name ***Root*** is said to *match* any name that starts with zero or more characters, then has the characters **Root**, then is followed by zero or more characters. As a result, *Mathematica* will show you all the names that contain the characters **Root**.

Browser-based Help

Version 2.x   In version 2.x of *Mathematica* for a graphically-based system, you can open up the interactive "Function Browser" from the Help menu. The upper-part of the browser window will look something like this:

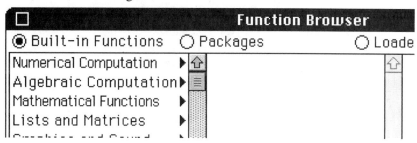

To look up information, say, about the **Solve** command, you check the list of categories and guess that the "Numerical Computation" category is a good place to start.

When you click on the "Numerical Computation" phrase, the second column will be filled in with a list of subcategories. Here, click on the "Equation Solving" phrase and you'll see a list of suitable commands appear in the third column.

When you click on "Solve" in the third column, the lower part of the browser window will be filled with information about the **Solve** command.

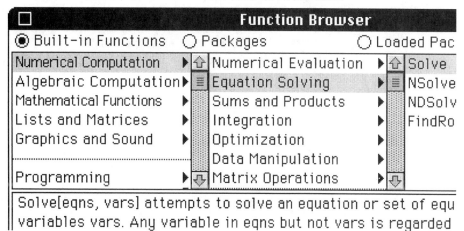

Version 3.0   In version 3.0, you launch the "Help Browser" by choosing the "Help..." item in the Help menu. This browser contains information about commands just like the version

2.x function browser. It also includes complete, on-line references to the *Mathematica Book*.

When you open it, its upper, left corner will look like this:

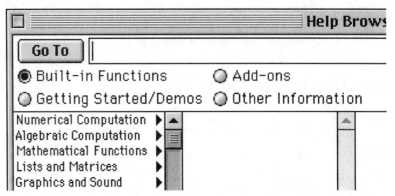

Its appearance is very similar to the version 2.x browser window. Notice that the same list of categories appears. Both sub- and sub-subcategories will appear as you click through the window.

There is one major addition to this window for version 3.0: the **Go To** button has a line next to it where you can type in a command or phrase. When you click the **Go To** button, *Mathematica* will go off and search for relevant information.

For example, to look up information on the **Solve** command, you type in the phrase "Solve" and click the **Go To** button:

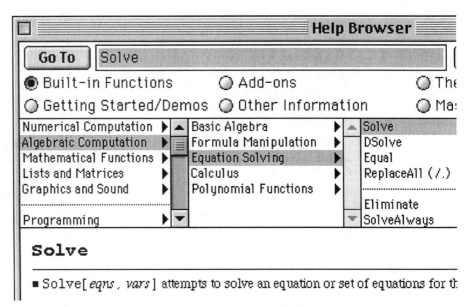

The information displayed comes straight from the *Mathematica* reference materials installed on your computer. It is much more informative than what you can get from asking **?Solve**.

Packages

Loading a Package

Mathematica has a built-in vocabulary of several hundred commands. But still these are not enough for everyday usage. Additional commands are available in the "Standard Packages." These include commands for Algebra, Calculus, Linear Algebra, Number Theory, Statistics, and so on.

Before using those commands, you have to load them using the **Needs** command. You have to specify the package by name with respect to where it is located in your packages folder (subdirectory).

For example, the standard package **DescriptiveStatistics** is located in a folder (subdirectory) called "Statistics." We will load it with

Needs["Statistics`DescriptiveStatistics`"]

> **Note:** Make sure that you use the double-quote " and the back-quote `
> when entering this. The back-quote ` usually shares the same key with
> the tilde character ~ at the top, left on most keyboards.

Now we can use the **Mean** and **Median** commands that are defined in this package:

Mean[{1, 5, 6, 9, 2}]   (*Gives the average of the numbers.*)

$$\frac{23}{5}$$

Median[{1, 5, 6, 9, 2}]   (*Gives the middle value.*)

5

Similarly, to use the **PieChart** command that's defined in the **Graphics** package inside the **Graphics** folder, you type:

Needs["Graphics`Graphics`"]
PieChart[{20, 10, 15, 5, 5}]

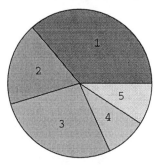

The complete list of commands for all standard packages can be found in the *Guide to Standard Packages* documentation that came with your software.

Help for Packages

You can use the Function Browser or the Help Browser to locate information about packages. Just click the "Package" button or the "Add-ons" button in the browser. Then you can work your way down the category lists to get information about the package you wanted.

Troubleshooting Q & A

Question... I was hoping to use the **Mean** command from the **DescriptiveStatistics** package, but I forgot to load it first:

> **Mean[{3,4,5}]**
>
> Mean[{3,4,5}]

Then I used the **Needs** command and got some error messages I didn't understand. What happened?

> **Needs["Statistics`DescriptiveStatistics`"]**
>
> Mean::shdw:
>   Symbol "Mean" appears in multiple contexts
>   "Statistics`DescriptiveStatistics`", "Global`"; definitions in
>   context "Statistics`DescriptiveStatistics`" may shadow or be
>   shadowed by other definitions."
>
> **Mean[{3,4,5}]**
>
> Mean[{3,4,5}]

Answer... This is one of the most confusing things about using packages. We'll be reminding you about this throughout the rest of this guide!

When you enter **Mean[{3,4,5}]** the first time, the **DescriptiveStatistics** package has not been loaded yet, so *Mathematica* thinks you're entering a new symbol named **Mean**. Then, when you load the package, a new definition for **Mean** is used.

Now you have two commands named **Mean**, and that's what the error message is telling you. To eliminate the confusion, you must explicitly ask *Mathematica* to **Remove** the symbol **Mean** that you accidentally defined first. Then, there'll be only one **Mean** left – the one defined in the package:

> **Remove[Mean]**
> **Mean[{3,4,5}]**
>
> 4

Question... My buddy told me that I can use the double less-than command to load a package like this:

> **<<Statistics`DescriptiveStatistics.m**

Is this a good idea?

Answer... No. There are two reasons.

- First, the << syntax depends on actual filenames in your file system. Package names may be different on your computer system. (On DOS systems, for example, filenames are limited to a certain number of characters.)

- Second, the **Needs** command is actually more subtle than just a "load a package" command. It only loads a package if it has not been loaded already. The << syntax can cause packages to be reloaded, and that can cause lots of confusion.

CHAPTER 8

Making Graphs

Drawing the Graph of a Function

The Plot Command

There are many ways that you can plot a two-dimensional picture in *Mathematica*. In this chapter we'll concentrate on drawing the graph of a function.

You draw graphs of functions $y = f(x)$ with the **Plot** command. To see the graph of $f(x) = x^2$ over the interval $-3 \le x \le 2$, type:

Plot[x^2, {x,-3,2}]

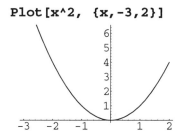

In general, to plot a function of x over an interval $a \le x \le b$, you type:

Plot[*function* , {x, *a*, *b* }]

Adjusting the PlotRange

When you use the **Plot** command, you don't need to specify the vertical range of the graph. *Mathematica* automatically adjusts the vertical range to show you the best part of the curve. However, beware!

Plot[x^3, {x,-2,2}]

Notice that a large portion of the graph above has been omitted! *Mathematica* displayed only the most interesting portion of the graph that's close to the origin. But you can use the **PlotRange** option for the **Plot** command to specify the vertical range that you want to see:

Plot[x^3, {x,-2,2}, PlotRange -> {-3,3}]

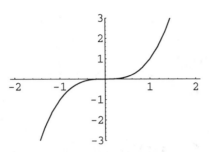

Now you see all *y*-values between –3 and 3. Using **Plot** with the **PlotRange** option is a lot like setting up a view window on a graphing calculator!

You still don't see the whole graph, however. You can force *Mathematica* to draw the entire graph with:

Plot[x^3, {x,-2,2}, PlotRange-> All]

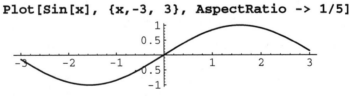

Proportion and AspectRatio   If you are using a graphically-based system, you can resize a picture by first clicking on it and then dragging along one of its corners or edges with a mouse.

Usually, no matter how big or small the picture you create, the **aspect ratio** of the picture (the ratio of the height to width) remains the same. The default aspect ratio is very close to the relative proportions of a standard credit card.

You can change the aspect ratio for a **Plot** command by specifying its **AspectRatio** directly. For example,

Plot[Sin[x], {x,-3, 3}, AspectRatio -> 1/5]

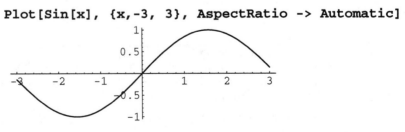

Here, the height of the picture is about one-fifth its width.

If you want the units on the *x*- and *y*-axes to be the same length, you must specify **AspectRatio** to be **Automatic**:

Plot[Sin[x], {x,-3, 3}, AspectRatio -> Automatic]

> **Note**: You should use **AspectRatio->Automatic** whenever you want to see angles or circles properly. For example, the graph of $f(x) = x$ will not look like it makes a 45° degree angle with the *x*-axis unless you specify **AspectRatio –> Automatic**. (See the pictures below.)

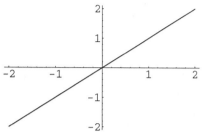   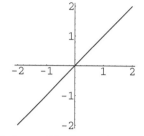

Without **AspectRatio –> Automatic**   *With* **AspectRatio –> Automatic**

Labeling a Picture

You can label a picture and its axes using the **PlotLabel** and **AxesLabel** options:

```
Plot[ Cos[x]+15, {x, 0,4*Pi},
        AxesLabel -> {"day", "price"},
        PlotLabel -> "Daily price of Stock AAA"]
```

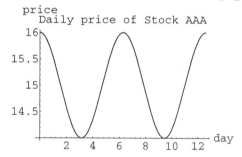

Tracing a Graph in Your Front End

In graphically-based systems you can find the coordinates of any point in a picture easily.

You first click the graphic with the mouse to select it. A frame will appear bounding the picture. Now hold down the **Apple** key (Macintosh) or the **Alt** key (DOS): the cursor changes to a cross-hair, and you'll see coordinates displayed at the bottom of your Notebook window. Move the mouse over the graph while still holding down the **Apple/Alt** key, and you can see the coordinates change correspondingly.

Plotting Multiple Expressions

Plotting Multiple Functions

You can plot several functions in the same picture by listing all the functions separated by commas, enclosing them in curly braces. For example:

```
Plot[{Sin[x], Cos[x]}, {x,-Pi,Pi}]
```

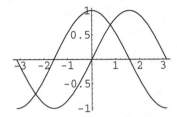

Combining Existing Plots with Show

You can also combine different graphics into a single picture by using the **Show** command. First, name each of the graphics. Then, use the **Show** command to combine these named graphics into a single output.

```
graph1 = Plot[Cos[x],{x,-3,1}]
graph2 = Plot[Sin[x],{x,0,3}]
graph3 = Plot[0.5,{x,-2,4}]
Show[graph1, graph2, graph3]
```

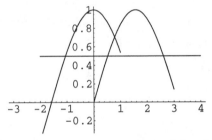

Mathematica will show you each of the pictures **graph1**, **graph2**, and **graph3** as you enter the commands. We only show the fourth (combined) picture above, which is the output of the **Show** command.

Style Options for Graphics

The PlotStyle Option

You can add more "character" to a graph by adjusting the color, thickness and dashing pattern of the plot. This can be done with the **PlotStyle** option inside the **Plot** command. It has the syntax:

```
Plot[ f(x), {x, a, b}, PlotStyle -> option]
```

or for multiple options:

```
Plot[ f(x), {x, a, b}, PlotStyle -> {{option1, option2, ..}}]
```

Thickness

In the following command, **Thickness[0.03]** means that the thickness of the pen will be about 3% of the width of the graphic.

```
Plot[ x^3+2x, {x, -3, 2},
        PlotStyle -> Thickness[0.03]]
```

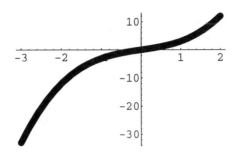

Dashing

In the following command, **Dashing[{0.02}]** means that about 2% of the curve will be drawn solid, then 2% omitted, then 2% drawn, then 2% omitted and so on. (Notice the curly braces { } inside the square brackets [ ].)

```
Plot[ x^3+2x, {x, -3, 2},
      PlotStyle -> Dashing[{0.02}]]
```

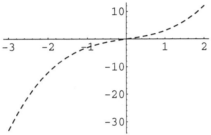

Colors

You can create any color by mixing the three primary colors red, green, and blue in various amounts. If you mix all three together at their full intensity, you get white. Mixing each at about half-intensity gives gray. You'll get a nice banana color by mixing red at 80% intensity, green at 80%, and blue at 30%.

The **RGBColor** directive lets you define colors by specifying the intensities of each of the red, green, and blue. For example, **RGBColor[0.8, 0.8, 0.3]** gives us our banana color:

```
Plot[ x^3+2x, {x, -3, 2},
      PlotStyle -> RGBColor[0.8, 0.8, 0.3] ]
```

Sorry, we cannot show you the picture here, because this text is printed in black and white!

If you want a colorful presentation, the following table lists some wild colors that you can try out:

Color	R-G-B Mixing	Color	R-G-B Mixing
Red	**RGBColor[1, 0, 0]**	LightBlue	**RGBColor[0.6, 0.8, 0.9]**
Blue	**RGBColor[0, 1, 0]**	Banana	**RGBColor[0.8, 0.8, 0.3]**
Green	**RGBColor[0, 0, 1]**	Carrot	**RGBColor[0.9, 0.5, 0.1]**
Magenta	**RGBColor[1, 0, 1]**	HotPink	**RGBColor[1, 0.4, 0.7]**
Chartreuse	**RGBColor[0.5, 1, 0]**	Raspberry	**RGBColor[0.5, 0.1, 0.3]**

**Multiple
PlotStyle
Options**

■ **Example.** Let us plot the sine and cosine curves in the same picture. To help distinguish them, we will draw them differently.

We'll make the sine curve show up in red with a dashed line and the cosine curve a thick blue. Here's how we define these style features, using the names **style1** and **style2**:

```
style1 = {RGBColor[1,0,0],Dashing[{0.02}]};
style2 = {RGBColor[0,0,1],Thickness[0.02]};
```

(*We used ";" at the end of each of the commands above to suppress the output.*)

```
Plot[{Sin[x],Cos[x]},{x,-Pi,Pi},
                    PlotStyle->{style1,style2}]
```

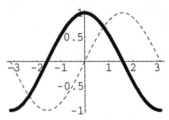

> **Note**: Be very careful when entering the **PlotStyle** options. A lot of curly braces, square brackets, and commas have to go in just the right places to make the **Plot** command work.

More Examples

Zooming In

■ **Example.** Consider $f(x) = \sqrt{1 + 10x^4 - 20x^5 + 25x^6}$ and $g(x) = x^2 + 5\sin(x)$:

```
f[x_] := Sqrt[ 1 + 10x^4 - 20x^5 + 25x^6]
g[x_] := x^2 + 5 Sin[x]
```

```
Plot[{f[x],g[x]},{x,-3,3}]
```

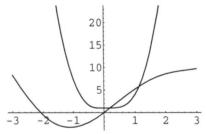

To get a better estimate of the intersection near 1.1, we can zoom in on the graph by successively shrinking down the *x*-interval in the **Plot** command.

```
Plot[{f[x],g[x]},{x,1,1.2}]
Plot[{f[x],g[x]},{x,1.13,1.15}]
Plot[{f[x],g[x]},{x,1.135,1.145}]
```

"Optical Illusion"

■ **Example.** Suppose we plot the function $f(x) = 64x^4 - 16x^3 + x^2$:

```
Plot[ 64x^4 - 16x^3 + x^2, {x, 0, 10}]
```

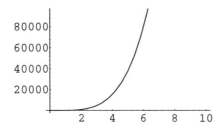

You can easily get the impression that the graph is always increasing. However, notice that the vertical range is very large, between 0 to 100,000, so the picture is not sharp enough to show any small dips. In particular, the graph looks like a straight line for x between 0 and 2; this certainly is not the case.

```
Plot[ 64 x^4 -16x^3+x^2, {x, 0, 2}]      (*Still looks OK.*)
Plot[ 64 x^4 -16x^3+x^2, {x, 0, 1}]      (*Still looks OK.*)
Plot[ 64 x^4 -16x^3+x^2, {x, 0, 0.3}]    (*Surprise!!*)
```

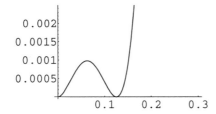

Useful Tips

💡 If you have a complicated **Plot** to draw, always try it first with no options to make sure it works. Add options for **Plot** only after you get a correct drawing.

Troubleshooting Q & A

Question... I got an empty picture from **Plot** with an error message "`. . . is not a machine-size real number.`" What went wrong?

Answer... This indicates that *Mathematica* cannot evaluate your input function numerically. Check whether you made a typo in the input. Common mistakes are:

- You forgot to use proper upper-case for a *Mathematica* built-in function.
- You typed the wrong name of a built-in function.
- You used the wrong variable.

Check whether your function really gives numbers! Do you get numbers when you enter **f[–1]**, **f[0]**, **f[1]**, and so on?

Question... I got the picture I wanted from **Plot**, but there is still an error message "`. . . is not a machine-size real number.`" What does that mean?

Answer... This means that your input function is not defined for every point in the specified interval. For example:

```
Plot[ 1/x, {x, 0, 2}]
```

The function $1/x$ is not defined at $x = 0$.

```
Plot[ Sqrt[1 - x^2], {x, 0, 2}]
```

The function $\sqrt{1 - x^2}$ gives imaginary numbers for $x > 1$. Of course, *Mathematica* cannot show you this part of the picture.

Question... When I combined various pictures using the **Show** command, I got an error message. Where is my mistake?

Answer... Check whether you typed the names of the pictures correctly in the **Show** command. For example:

```
Show[pict1, pict2, pict3, pict4]
```

```
An error was encountered in combining the graphics
objects in Show[-Graphics-,pict2,pict3,-Graphics-].
```

This error message tells you that *Mathematica* has trouble with **pict2** and **pict3**. They do not appear to be graphics. Recheck their definitions.

Question... When I used various options in the **Plot** command, I got the error message "`An option must be a rule or a list of rules.`" What went wrong?

Answer... A common mistake is to type the equal sign "=" instead of the arrow "–>". Also, when you enter the arrow "–>", make sure that there is no space between the minus sign "–" and the greater-than sign ">".

Question... I tried to **Plot** a picture with two or more **PlotStyle** options, but only one of the options showed up in the graph. What went wrong?

Answer... You have to input multiple **PlotStyle** options for a single graph with the following format, using *two* pairs of curly braces:

```
Plot[ f(x), {x,a,b}, PlotStyle -> {{option1, option2, ..}}]
```

For example, *Mathematica* will only draw a thick line without coloring if you type:

```
Plot[x, {x,0, 1}, PlotStyle ->
        { Thickness[0.02], RGBColor[0.6,0.8,0.2]} ]
```

You have to type the following to see the color:

```
Plot[x, {x,0, 1}, PlotStyle ->
        {{ Thickness[0.02], RGBColor[0.6,0.8,0.2]}} ]
```

CHAPTER 9
Plotting a Curve

Parametric Plots

Plotting Parametric Curves

In the previous chapter you saw how to use **Plot** to draw curves that are graphs of functions. But not all curves are the graphs of functions.

A two-dimensional (2-D) parametric curve is written in the form $(x(t), y(t))$. You use the **ParametricPlot** command to draw it over an interval $a \le t \le b$. The command has the form:

```
ParametricPlot[{ x(t), y(t)}, { t, a, b }]
```

For example, to see the curve $(t^2 - 1, \ t + t^2)$ for $-2 \le t \le 2$, we use:

```
ParametricPlot[{t^2-1,t+t^2}, {t,-2,2}]
```

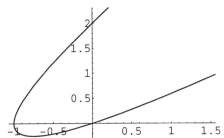

AspectRatio Option

Most of the options you can use with the **Plot** command (e.g., **AspectRatio**) work with the **ParametricPlot** command as well.

■ **Example.** To see the unit circle $(\cos t, \sin t)$ for $0 \le t \le 2\pi$, you use:

```
ParametricPlot[{Cos[t],Sin[t]},{t,0,2Pi}]
```

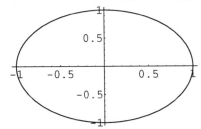

This does not look like a circle because the two axes are not measured in the same unit length. To correct this, we use the option **AspectRatio –> Automatic** just as we did in the **Plot** command.

```
ParametricPlot[{Cos[t],Sin[t]},{t,0,2Pi},
   AspectRatio->Automatic]
```

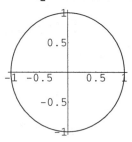

Multiple Curves

The **ParametricPlot** command lets you plot several curves in one picture, just like the **Plot** command. To do this, you put all the expressions of the curves inside a list.

■ **Example**. To draw the two curves (t^2, t) and (t, t^2), for $-2 \le t \le 2$, use:

```
ParametricPlot[{ {t^2,t},{t,t^2} }, {t,-2,2}]
```

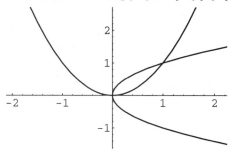

Multiple Styles

You can specify different characteristics for each of the curves you draw, just like you do in the **Plot** command.

■ **Example**. Consider the following two curves:

$$(t + \sin(t), 1 + \cos(t)) \text{ and } (t + \sin(t + \pi), 1 + \cos(t + \pi)) \text{ for } 0 \le t \le 4\pi.$$

(These are called cycloids. When a circle of radius 1 is "rolling" along the x-axis, the curve traced by a fixed point on the circumference is called a cycloid. We have an animation of the cycloid in Chapter 23.)

We will first input the expressions of the x- and y-coordinates:

```
x1[t_]:=t+Sin[t];      y1[t_]:=1+Cos[t];
x2[t_]:=t+Sin[t+Pi];   y2[t_]:=1+Cos[t+Pi];
```

To make the first curve appear in gray and the second with some dashing, you use the same **PlotStyle** options as in a **Plot** command:

```
style1 = {GrayLevel[0.5]};           (*Define styles.*)
style2 = {Dashing[{0.05}]};

ParametricPlot[
   {{x1[t],y1[t]},{x2[t],y2[t]}},{t,0,4Pi},
   PlotStyle->{style1,style2},
   AspectRatio->Automatic]
```

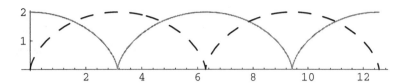

Multiple Pictures Together

You can combine several 2-D pictures with the **Show** command. The pictures you combine can even be created by completely different graphic commands such as **Plot**, **ListPlot**, and **ParametricPlot**.

■ **Example**. To create the "open skull" picture that you see below, we will combine several pictures. The face is made from a parabola and two straight lines; the eyes from two circles; the nose by connecting three vertices of a triangle; and the mouth from a straight line.

```
face = Plot[{x^2, 4, 4.5+x/4},{x,-2,2}]
eyes = ParametricPlot[{{1+Cos[t]/2,3+Sin[t]/2},
          {-1+Cos[t]/2, 3+Sin[t]/2}},{t,0,2Pi}]
nose = ListPlot[{{-0.5,2},{0.5,2},{0,2.5},
          {-0.5,2}}, PlotJoined ->True]
mouth = ListPlot[{{-0.5,1},{0.5,1}},
          PlotJoined -> True]

Show[face, eyes, nose, mouth, PlotRange -> All,
     AspectRatio -> Automatic]
```

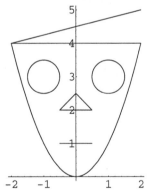

Troubleshooting Q & A

Question... I tried to use the **ParametricPlot** command, but *Mathematica* just gave the command back to me and complained about "Possible spelling error." What happened?

Answer... Check whether you spelled the word **ParametricPlot** correctly. Make sure that you used the upper case character for both letters "**P**."

Question... When I used **ParametricPlot**, I got an error message "Invalid input." What should I look for?

Answer... Check if you entered the syntax correctly. Also a common mistake for beginners is to enter, for example, **(t, t^2)** instead of **{t, t^2}** for the curve (t, t^2).

Question... When I used **ParametricPlot**, I got the error message "does not evaluate to a pair of real numbers." What happened?

Answer... This indicates that *Mathematica* cannot evaluate your input function numerically. Check whether you made a typo in the input. Common mistakes are:

- You forgot to use proper upper case for a *Mathematica* built-in function.
- You used a wrong parameter.

Question... When I used **ParametricPlot**, I got the message "Function ... cannot be compiled." I still got the picture I wanted, though. Should I worry about this?

Answer... No. **ParametricPlot** is fussy about the form of the parametric definition you give for your curve. If it is a list of two functions, *Mathematica* rewrites the functions internally in an efficient form which speeds up the command. This process is called *compilation*.

This happens, for example, with:

```
ParametricPlot[ {t,t^2}, {t,0,2} ]
```

However, if you don't list the functions explicitly, *Mathematica* cannot speed up computation using compilation. This would happen, for example, with:

```
r[t_] := {t,t^2}
ParametricPlot[ r[t], {t,0,2} ]
```

This generates the message above, but it's only a warning. And you won't notice any real speed difference. Relax, and enjoy the picture!

CHAPTER 10

PolarPlot and ImplicitPlot

Plotting in Polar Coordinates

The PolarPlot Command

If a curve is expressed in polar coordinates, then we can draw it with the **PolarPlot** command. This command is defined in an external package. (See Chapter 7 about packages.) You first load it with:

```
Needs["Graphics`Graphics`"]
```

(Watch carefully for the double-quote " and the back-quote ` when entering this. The back-quote ` usually shares the same key with the tilde character ~ at the top, left on most keyboards.)

If the curve is given by $r = f(\theta)$, for $\theta_1 \le \theta \le \theta_2$, we can draw it with the following.

```
PolarPlot[ f(θ) , { theta, θ₁ , θ₂ }]
```

Notice that this form is very similar to that of the **Plot** command you already know. For example, to plot the three-leaf rose $r = 2\cos(3\theta)$, use:

```
Needs["Graphics`Graphics`"]     (*If you haven't already loaded
                                    the package.*)
PolarPlot[ 2*Cos[3*theta] ,{theta, 0, 2Pi}]
```

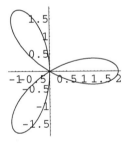

Plotting Multiple Curves

Just as in the **Plot** or **ParametricPlot** command, you can plot more than one curve at the same time by entering all of them using a list.

For example, you can plot the spirals $r = \frac{\theta}{2\pi}$, $r = \left(\frac{\theta}{2\pi}\right)^2$, and the circle $r = 1$ all in one picture with:

```
PolarPlot[{ theta/(2*Pi), (theta/(2*Pi))^2, 1 },
     {theta, 0, 2Pi}]
```

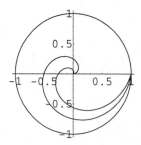

> **Note:** **AspectRatio –> Automatic** is the default setting for **PolarPlot**. Some polar curves will have poor default proportions, so you may have to set **AspectRatio** explicitly to get a good picture.

Controlling Styles

All the **PlotStyle** options that can be used with **Plot** also work with the **PolarPlot** command.

■ **Example**. To see the two cardioids $r = 1 + \cos\theta$ and $r = 1 - \sin\theta$, you can use the following syntax.

```
style1 = {Thickness[0.015],GrayLevel[0.5]};
style2 = {Dashing[{0.05,0.02}]};

PolarPlot[{1 + Cos[theta], 1 - Sin[theta]},
                {theta, 0, 2Pi},
                PlotStyle->{style1,style2}]
```

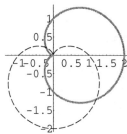

Plotting Graphs of Equations

Implicit Plots

When a curve is given by an equation in variables x and y, you can draw the curve with the **ImplicitPlot** command. You have to load the package that defines it first, however. Again, watch carefully for the double-quote " and the back-quote ` when entering this:

```
Needs["Graphics`ImplicitPlot`"]
```

You use the command in the form:

> **ImplicitPlot[** *an equation in x and y* **, { x,** $x_{\min}$ **,** $x_{\max}$ **}]**

For example, to see the unit circle $x^2 + y^2 = 1$ for $-1 \le x \le 1$, use:

```
Needs["Graphics`ImplicitPlot`"]    (*If you haven't already
                                      loaded the package.*)
ImplicitPlot[ x^2 + y^2 == 1, {x, -1, 1} ]
```

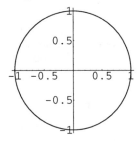

> **Notes**
>
> (1) Make sure you use double equals "==" when you enter the equation.
>
> (2) **AspectRatio –> Automatic** is the default setting for **ImplicitPlot**.

Multiple Curves and Styles

You can plot more than one implicit curve in a single command. You can specify different styles for each curve as well. This is similar to the way you do it in the **Plot**, **ParametricPlot**, and **PolarPlot** commands.

■ **Example**. We want to see the curves $x^2 + 3xy + y^3 = 25$, $x^2 + 3xy + y^3 = 10$ and $x^2 + 3xy + y^3 = 0$ in the interval $-10 \le x \le 10$. Let's draw each curve with a different color:

```
f[x_,y_] := x^2 + 3x*y + y^3
style1 = { RGBColor[0.8, 0.8, 0.2] };
style2 = { RGBColor[0.6, 0.2, 0.4] };
style3 = { RGBColor[0.2, 0.5, 0.8] };
ImplicitPlot[
   {f[x,y] == 25, f[x,y] == 10, f[x,y] == 0},
   {x,-10,10}, PlotStyle -> {style1, style2, style3},
   AspectRatio -> 1/GoldenRatio ]
```

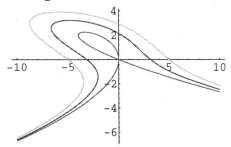

This picture looks like the Chinese character for "Wind."

Useful Tips

 You may be overwhelmed by all the different plotting commands that we have discussed. Don't worry, **Plot** and **ParametricPlot** are the most commonly used commands, while **PolarPlot** and **ImplicitPlot** are for special situations:

To Draw	Use
Graph of a function $f(x)$	`Plot`
Parametric curve $(x(t), y(t))$	`ParametricPlot`
Curve in polar coordinates	`PolarPlot`
Curve given by an equation	`ImplicitPlot`

 Instead of loading the packages **Graphics`Graphics`** and **Graphics`ImplicitPlot`** separately, you can enter the single command:

    Needs["Graphics`Master`"]

This command will load the names of all commands defined in the **Graphics** packages. This way, you don't have to remember the specific name of the package that defines **PolarPlot** or **ImplicitPlot**.

Troubleshooting Q & A

Question... I tried to use the **PolarPlot** command, but *Mathematica* just gave the command back to me without doing anything. What happened?

Answer... Usually, there are two reasons why this might happen:

- Check whether you spelled the word **PolarPlot** correctly. Make sure that you used an upper-case character for both letters "**P**."

- **PolarPlot** is defined in an external package and must be loaded before you can use it. If you try to use it without loading it, *Mathematica* doesn't have a definition for **PolarPlot**, so it can't do anything with it. To correct this problem, use the following sequence and then try the command again:

      Remove[PolarPlot]
      Needs["Graphics`Graphics`"]

Question... I got an empty graphic when I used **PolarPlot**. What happened?

Answer... Check whether you made a typo in the input. Common mistakes are:

- You forgot to use upper case for a *Mathematica* built-in function.

- Your input function contains a variable other than **theta** (a common mistake is to include **r**, the radius variable, in the input).

- You did not match up parameters correctly (e.g., you used **t** in the function, but used **theta** when you specified the interval).

Question... I tried to use the **ImplicitPlot** command, but *Mathematica* just gave the command back to me without doing anything. What happened?

Answer... **ImplicitPlot** is also defined in an external package, just like **PolarPlot**, and must be loaded before you can use it. To correct the problem, use this sequence first and then try the **ImplicitPlot** command again:

```
Remove[ImplicitPlot]
Needs["Graphics`ImplicitPlot`"]
```

Question... **ImplicitPlot** gave me an error message I don't understand. What should I look for?

Answer... There are two major problem areas in using **ImplicitPlot**.

- Make sure that the equation you entered is really an equation that uses the double-equal sign "==", and not the equal sign "=". If you used the equal sign, you've written an assignment statement that probably doesn't make much sense.

- Make sure that your equation has two variables that do not have values. Executing **Clear[x,y]** before using **x** and **y** in your equation for **ImplicitPlot** is highly recommended.

Question... I got some very strange pictures from **ImplicitPlot** that were either very wide or very tall. What can I do?

Answer... Most curves you draw with **ImplicitPlot** will have poor default proportions, so you may have to set **AspectRatio** explicitly to get a good picture. We suggest you draw the graphic by directly setting the option **AspectRatio –> 1/GoldenRatio**. (This is *Mathematica*'s default aspect ratio.)

Question... I got an empty or partial graphic when I used **ImplicitPlot**. What happened?

Answer... This will happen if the equation you give cannot be solved meaningfully in terms of the variable you specify. For example, in the command

```
ImplicitPlot[ x^2-x*y==1, {x, -5, 5}]
```

Mathematica first rewrites $x^2 - xy = 1$ as $y = (1 - x^2)/(-x) = x - 1/x$. Then it tries to plot y-values using $y = x - 1/x$ for many x-values with $-5 \le x \le 5$. When $x = 0$, the equation is meaningless and **ImplicitPlot** just stops working.

ImplicitPlot can sketch a better picture, if you specify intervals for *both* the x- and y-variables, like the following:

```
ImplicitPlot[x^2-x*y==1, {x,-5,5}, {y,-5,5},
          AspectRatio->1/GoldenRatio]
```

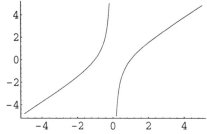

CHAPTER 11

Limits and Derivatives

Limits

The Limit Command

If f is a function of a single variable, *Mathematica* evaluates $\lim_{x \to a} f(x)$ with the following syntax:

Limit[*function,* **x -> ** *a***]**

For example, *Mathematica* agrees that $\lim_{x \to 0} \dfrac{\sin x}{x} = 1$:

Limit[Sin[x]/x, x -> 0]

1

The following table shows some sample limit computations. *Mathematica* can compute most limits, even those that involve infinite limits, or limits at infinity:

Limit Calculation	Mathematica *Evaluation*
$\lim_{x \to 0} \dfrac{e^x - 1 - x}{x^2} = \dfrac{1}{2}$	**Limit[(Exp[x]-1-x)/x^2,x -> 0]** $\dfrac{1}{2}$
$\lim_{x \to 3} \dfrac{1-x}{(x-3)^2} = -\infty$	**Limit[(1-x)/(x-3)^2,x -> 3]** -Infinity
$\lim_{x \to +\infty} \dfrac{x}{\sqrt{x^2+1}} = +1$	**f[x_] := x/Sqrt[x^2+1]** **Limit[f[x],x -> Infinity]** 1
$\lim_{x \to -\infty} \dfrac{x}{\sqrt{x^2+1}} = -1$	**Limit[f[x],x -> -Infinity]** -1

If *Mathematica* returns a limit expression unevaluated, then either the limit does not exist or *Mathematica* cannot determine its value. For example, in *Mathematica* 2.2 (or a previous version):

Limit[Abs[x] / x, x-> 0]

$\text{Limit}[\dfrac{\text{Abs}[x]}{x}, x->0]$

Limit[Abs[x] / x, x-> 1]

$\text{Limit}[\dfrac{\text{Abs}[x]}{x}, x->1]$

even though $\lim\limits_{x\to 0} |x|/x$ does not exist and $\lim\limits_{x\to 1} |x|/x = 1$. (*Mathematica* 3.0 will give different answers.)

One-sided Limits

To calculate a left-hand limit $\lim\limits_{x\to a^-} f(x)$, you have to specify the **Direction** option in the **Limit** command. To find $\lim\limits_{x\to 0^-} \sqrt{2x^2}\,/x$, you use:

```
Limit[Sqrt[2x^2]/x, x -> 0, Direction -> 1]
```
$-\sqrt{2}$

Direction is set to **1** because, in this left-hand limit, x is increasing (in the *positive* direction) as it approaches 0.

Similarly, to find the right-hand limit $\lim\limits_{x\to 0^+} \sqrt{2x^2}\,/x$, you use:

```
Limit[Sqrt[2x^2]/x, x -> 0, Direction -> -1]
```
$\sqrt{2}$

Now, x approaches 0 from the right and is decreasing (in the *negative* direction), thus we use **Direction -> -1.**

> **Note:** The + or – sign you use for the **Direction** is exactly opposite to the sign in $x \to a^-$ or a^+ you use to denote the limit mathematically.

If you don't specify **Direction**, *Mathematica* uses **Direction -> -1** automatically. That is, **Limit[f[x], x -> a]** actually gives you the right-hand limit $\lim\limits_{x\to a^+} f(x)$.

Differentiation

Differentiation Using Prime Syntax

If f is a function of a single variable, *Mathematica* will understand the symbol f' as the derivative of f, where the prime is entered using the single-quote character. For example,

```
Clear[f]
f[x_] := x^4
f'[x]
```
$4\,x^3$

Higher-order derivatives follow with the usual notation:

```
f''[x]
```
$12\,x^2$

```
f'''[a]
```
$24\,a$

Differentiation Using D	Also, you can use the differential operator **D** for computing derivatives.

> **D[** *function* **,** *variable* **]**

For example,

```
f[x_] := x^4
D[f[x],x]
```

$4\,x^3$

To calculate a second derivative, you can use **D** twice via composition.

```
D[D[f[x],x],x]
```

$12\,x^2$

Or you can use either of these short-hand formats:

```
D[f[x], x, x]
```

$12\,x^2$

```
D[f[x], {x,2}]
```

$12\,x^2$

Similarly, the third derivative of *f* can be given by either of the following:

```
D[f[x], x, x, x]
```

$24\,x$

```
D[f[x], {x,3}]
```

$24\,x$

A few other examples of derivatives are given in the following table.

Mathematical Expression	Mathematica *Evaluation*
$\dfrac{d(x^2+e^{x^3})}{dx}$	`D[x^2+Exp[x^3], x]` $2\,x + 3\,E^{x^3}\,x^2$
$\dfrac{d(\cos(y^2)+y^5)}{dy}$	`D[Cos[y^2] + y^5, y]` $5\,y^4 - 2\,y\,\mathrm{Sin}\!\left[y^2\right]$
$\left.\dfrac{df}{dt}\right|_{t=1}$, where $f(t)=t^2+t^3+\ln(t)$	`f[t_] := t^2 + t^3 + Log[t]` `f'[1]` 6 *or, we could use this substitution expression:* `D[f[t],t] /. {t -> 1}` 6

Differentiation Rules	*Mathematica* knows all the formal computation rules in differentiation, such as

```
Clear[f, g, h]
D[f[x]*g[x], x]                (*The product rule*)
```

$g[x]\,f'[x] + f[x]\,g'[x]$

> **D[f[x]/g[x], x]** (*The quotient rule*)
>
> $$\frac{f'[x]}{g[x]} - \frac{f[x] \, g'[x]}{g[x]^2}$$
>
> **D[f[x]^n, x]** (*The power rule*)
>
> $$n \, f[x]^{-1+n} \, f'[x]$$
>
> **D[f[g[x]], x]** (*The chain rule*)
>
> $$f'[g[x]] \, g'[x]$$

How about the rule for differentiating the product of three functions?

> **D[f[x]*g[x]*h[x], x]**
>
> $$g[x] \, h[x] \, f'[x] + f[x] \, h[x] \, g'[x] + f[x] \, g[x] \, h'[x]$$

More Examples

Limits and Graphs

■ **Example.** We can compute $\displaystyle\lim_{x\to 0} \frac{\sin(\sin^2(2x))}{x^2}$ with:

> **Limit[Sin[Sin[2x]^2]/x^2, x -> 0]**
>
> 4

We can see this answer both numerically and graphically.

- (*Numerically*) We substitute various test values of x that are close to zero and check whether the values of the function approach 4:

> **Sin[Sin[2x]^2]/x^2 /.**
> **   x -> {0.12, -0.1, 0.0123, -0.0125, 0.001234}**
>
> { 3.9217, 3.94593, 3.99919, 3.99917, 3.99999 }

- (*Graphically*) We see that the height of the function approaches 4, as x approaches zero.

> **Plot[Sin[Sin[2x]^2]/x^2, {x, -0.5, 0.5}]**

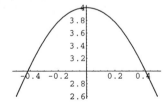

Definition of Derivative

■ **Example.** Consider the function $f(x) = x^2 \sin(x) + \cos(x)$:

> **f[x_] := x^2*Sin[x] + Cos[x]**

According to the definition of derivative, we have $f'(x) = \displaystyle\lim_{h\to 0} (f(x+h) - f(x))/h$.

So when h is sufficiently small, say $h = 0.1$, we would expect $(f(x+0.1) - f(x))/0.1$

to be very close to $f'(x)$. We can see this by plotting these two functions on the same graph:

```
Plot[{f'[x], (f[x+0.1]-f[x])/0.1}, {x,-3,3}]
```

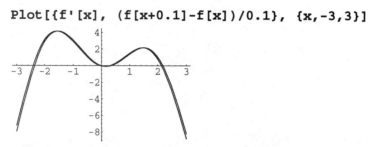

The result can be even better if we choose $h = 0.01$:

```
Plot[{f'[x], (f[x+0.01]-f[x])/0.01}, {x,-3,3}]
```

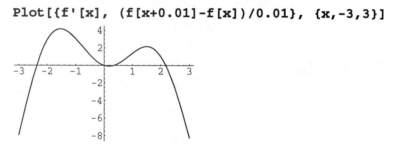

The Geometry of the Derivative

■ **Example.** Consider the function $f(x) = 3x^2 - 6x\cos(x)$:

```
f[x_] := 3x^2 - 6x*Cos[x]
```

The value of the derivative at $x = a$, $f'(a)$, gives the slope of the tangent line at that point. Using this result, we know that the equation of the tangent line is given by

$$y = f(a) + f'(a)(x - a).$$

We can check this by combining the graph of $y = f(x)$ with the tangent lines at several x-values, say at $x = -2.7$, $x = -1.4$ and $x = 0.9$.

```
graph1 = Plot[f[x], {x,-3.5,2},
              PlotStyle -> {{GrayLevel[0.5]}}]
line1 = Plot[f[-2.7]+f'[-2.7]*(x+2.7), {x,-3.5,-2}]
line2 = Plot[f[-1.4]+f'[-1.4]*(x+1.4), {x,-2,0}]
line3 = Plot[f[0.90]+f'[0.90]*(x-0.90),{x,0,2}]

Show[graph1, line1, line2, line3]
```

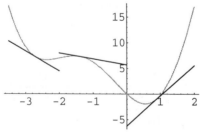

Useful Tips

 We recommend that you use **f '[x]** rather than **D[f[x], x]** for the first derivative $f'(x)$. **f '[x]** is simpler to write. Also, you'll run into fewer of the tricky syntax problems of *Mathematica* when you use **f '[x]** with commands such as **Plot**.

Troubleshooting Q & A

Question... *Mathematica* returned my limit expression unevaluated. Why did it do that?

Answer... This means that either the limit does not exist or *Mathematica* cannot determine its value. Recheck the examples we gave at the start of this chapter.

Question... I got zero for the derivative of a function, yet I know that the derivative is not zero. What should I check?

Answer... You should check the following:

- Make sure that you are differentiating with respect to the correct variable. For example, you may have defined a function in terms of t but differentiated with respect to x.

- A common mistake is to type **D[f, x]** to find the derivative, df/dx. This is wrong, because this command means differentiation of the *variable f* and not the *expression* $f(x)$. You have to type **D[f[x], x]**.

Question... I had lots of trouble defining a new function that used the derivative command **D**. What should I look for?

Answer... The usual way of defining a new function using **:=** doesn't work very well with the **D** command. For example, if you try

```
g[x_] := D[x^2, x]
g[x]

2 x
```

It looks OK, but when you ask, say,

```
g[2]

General::"ivar": "2 is not a valid variable.
```

This is wrong because *Mathematica* is trying to calculate **g[2] = D[2^2, 2]** which does not make sense. The correct method is to define g with **=** instead, using immediate assignment.

```
Clear[g,x]
g[x_] = D[x^2, x]

2 x
```

```
g[2]
```
4

Question... I can't seem to get a plot of the derivative of a function.

Answer... The command

```
Plot[ D[f[x], x], {x, -2, 2}]
```

has the same trouble that we pointed out in the previous answer. What happens is that *Mathematica* forms the expression **D[f[x],x]** at several x-values *first* and *then* tries to evaluate **D**. For example, to plot a value at $x = -2$, **Plot** tries to figure out what **D[f[-2], -2]** means.

The problem here is more a quirk of the **Plot** command than anything having to do with the derivative **D**. To make it work, you have to force the derivative to be evaluated first before letting **Plot** evaluate it. You do this either with:

```
Plot[ Evaluate[ D[f[x], x]], {x, -2, 2}]
```

or with:

```
Plot[ f'[x], {x, -2, 2}]
```

CHAPTER 12
Integration

Anti-Differentiation

The Integrate Command

You can use the **Integrate** command to compute $\int f(x)\,dx$. It has the form:

> **Integrate[** $f(x)$**, x]**

You specify a function or an expression to integrate, as well as the variable in which the integration is to take place.

■ **Example.** To compute $\int x^2\,dx$, use this syntax:

> **Integrate[x^2, x]**
>
> $$\frac{x^3}{3}$$

Note: When *Mathematica* does integration, it may be very slow in responding. The first integration you do will usually take longer, too.

Mathematica can integrate almost every integral that can be done using standard integration methods (e.g., substitution, integration by parts, partial fractions). Here are some typical integrations:

Integral	*In* Mathematica	*Comment*
$\int x\ln(x)\,dx$	**Integrate[x*Log[x], x]** $$-\frac{x^2}{4} + \frac{1}{2}\,x^2\,\text{Log[x]}$$	This one uses integration by parts!
$\int \dfrac{y^2}{\sqrt{1-y^2}}\,dy$	**Integrate[y^2/Sqrt[1-y^2], y]** $$-\frac{1}{2}\,y\,\sqrt{1-y^2} + \frac{\text{ArcSin[y]}}{2}$$	This is computed using a trigonometric substitution (write $y = \sin u$).
$\int \sin(\cos(y^2))\,dy$	**Integrate[Sin[Cos[y^2]],y]** $$\int \text{Sin}\left[\text{Cos}\left[y^2\right]\right]\,dy$$	This integrand has no closed-form anti-derivative.
$\int \dfrac{x}{\sqrt{1-x^4}}\,dx$	**Integrate[x/Sqrt[1-x^4], x]** $$\frac{\text{ArcSin}\left[x^2\right]}{2}$$	*Mathematica 2.2* cannot do this one, but version 3.0 can!

When *Mathematica* can't handle an integral, it usually returns your input unevaluated. This can mean either that it's not possible to find an antiderivative in closed form or that *Mathematica* hasn't yet been programmed to do the integral.

Definite Integrals

A definite integral $\int_a^b f(x)\,dx$ is computed in *Mathematica* with this form of the **Integrate** command:

> **Integrate[** $f(x)$, **{x, a, b}]**

Mathematica will try to find an antiderivative first, then evaluate it at the endpoints and subtract (according to the Fundamental Theorem of Calculus). Here are some examples:

Integral	In Mathematica	Comment
$\int_1^2 x^2\,dx$	`Integrate[x^2, {x,1,2}]` $\dfrac{7}{3}$	$\left.\dfrac{x^3}{3}\right\|_1^2 = \dfrac{8}{3} - \dfrac{1}{3} = \dfrac{7}{3}$
$\int_2^\infty \dfrac{1}{4+t^2}\,dt$	`Integrate[1/(4+t^2),{t,2,Infinity}]` $\dfrac{\pi}{8}$	This is an improper integral that involves $+\infty$.
$\int_0^1 \sqrt{\cos(x^2)}\,dx$	`Integrate[Sqrt[Cos[x^2]], {x,0,1}]` $\int_0^1 \sqrt{\cos[x^2]}\,dx$	There's no anti-derivative. *Mathematica* can't evaluate it at the endpoints.

Numerical Integration

The NIntegrate Command

We can use the **NIntegrate** command to find a numerical approximation for the integral $\int_a^b f(x)\,dx$ with the following syntax:

> **NIntegrate[** $f(x)$, **{x, a, b}]**

Notice that its syntax is the same as the **Integrate** command.

■ **Example.** To get approximate, numerical values for the integrals $\int_0^2 x^2\,dx$ and $\int_0^1 \sqrt{\cos(x^2)}\,dx$, use these commands:

```
NIntegrate[x^2, {x,0,2}]
2.66667

NIntegrate[Sqrt[Cos[x^2]], {x,0,1}]
0.948522
```

The **NIntegrate** command doesn't attempt to find a symbolic antiderivative, so it's quick, and it works with almost all integrands, including $\int_0^1 \sqrt{\cos(x^2)}\, dx$ for which **Integrate** failed (as you saw above).

More Examples

Area Between Curves

■ **Example.** To approximate the area bounded by the curves $p(x) = x^5 - 20x^3$ and $q(x) = 30 - x^5$, we start by sketching the curves. First, let's see where they intersect:

```
Clear[p,q]
p[x_] := x^5 - 20x^3
q[x_] := 30 - x^5
```

```
NSolve[ p[x] == q[x], x]
{{x -> -3.08004}, {x -> -1.20632},
   {x -> 0.527291 - 0.985474 I},
   {x -> 0.527291 + 0.985474 I}, {x -> 3.23178}}
```

The real values –3.08, –1.206 and 3.232 give approximations for the x-coordinates of the intersections. We can highlight the area between these two curves using the **FilledPlot** operator:

```
Needs["Graphics`FilledPlot`"]
FilledPlot[ {p[x], q[x]}, {x, -3.1, 3.3},
            PlotRange -> All]
```

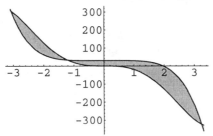

The area can be approximated with $\int_{-3.08}^{3.232} |p(x) - q(x)|\, dx$ using **NIntegrate**:

```
NIntegrate[Abs[p[x]-q[x]], {x, -3.08,3.232}]
388.854
```

The area can also be approximated as $\int_{-3.08}^{-1.206} p(x) - q(x)\, dx + \int_{-1.206}^{3.232} q(x) - p(x)\, dx$:

```
Integrate[p[x]-q[x], {x,-3.08,-1.206}] +
    Integrate[q[x]-p[x], {x,-1.206,3.232}]
388.854
```

Fundamental Theorem of Calculus

■ **Example.** The Fundamental Theorem of Calculus states that if f is a continuous function, then $\dfrac{d}{dx}\displaystyle\int f(x)\,dx \;=\; f(x)$. So if you integrate a function f and then differentiate the result, you should expect to get f back again. Let's try it:

```
Integrate[1/(x^3+1),x]
```

$$\frac{\text{ArcTan}\left[\frac{-1+2x}{\sqrt{3}}\right]}{\sqrt{3}} + \frac{1}{3}\,\text{Log}[1+x] - \frac{1}{6}\,\text{Log}[1-x+x^2]$$

```
D[%,x]
```

$$\frac{1}{3\,(1+x)} - \frac{-1+2\,x}{6\,(1-x+x^2)} + \frac{2}{3\left(1+\frac{1}{3}\,(-1+2\,x)^2\right)}$$

It looks like *Mathematica* messed up! But wait ...

```
Simplify[%]
```

$$\frac{1}{1+x^3}$$

Plotting an Antiderivative

■ **Example.** *Mathematica* cannot find an explicit formula for the antiderivative $\displaystyle\int\sqrt{1.1+\cos(10x^2)}\,dx$. However, you can still plot its graph with the help of the **NIntegrate** command.

The Fundamental Theorem of Calculus says that $F(x) = \displaystyle\int_0^x\sqrt{1.1+\cos(10t^2)}\,dt$ is an antiderivative of $\sqrt{1.1+\cos(10x^2)}$. *Mathematica* can compute values of F using **NIntegrate**, and you can **Plot** these values.

```
Plot[NIntegrate[ Sqrt[1.1+Cos[10t^2]], {t,0,x}],
   {x,0,1}]
```

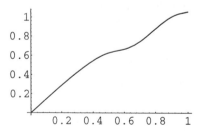

Useful Tips

💡 💡 Always use **NIntegrate** to evaluate a definite integral, unless you need an exact answer. In many cases, **Integrate** will neither work nor give you a useful result. Even when **Integrate** does work, it can be very slow.

Troubleshooting Q & A

Question... I got some strange answers for definite integrals that used names like "Erf," "FresnelS," and "FresnelC." What happened?

Answer... *Mathematica* knows about a lot of **special functions** that appear quite often in integration problems. For example, "Erf" is the *Mathematica* name for

$$erf(x) = \frac{2}{\sqrt{\pi}} \int_0^x e^{-t^2} \, dt.$$

This is an integral that does not have a closed-form antiderivative, but its values are well known. You might see something like this:

```
Integrate[ Exp[-t^2], {t,0,1} ]
```

$$\frac{1}{2} \sqrt{\pi} \, \text{Erf}[1]$$

```
N[%]
```

0.746824

Mathematica has reported $\int_0^1 e^{-t^2} \, dt = \frac{\sqrt{\pi}}{2} (\frac{2}{\sqrt{\pi}} \int_0^1 e^{-t^2} \, dt) = (\frac{\sqrt{\pi}}{2}) erf(1) \approx 0.747.$

Question... When I used **NIntegrate**, I got the error message "Integrand ... is not numerical." What should I check?

Answer... Check whether you made an error in the input. Also check if you used more than one variable in the input. For example, if the variable **a** is unassigned, you can do

```
Integrate[ x^2 + a^2, {x, 0, 1}]
```

but will get an error message with

```
NIntegrate[ x^2 + a^2, {x, 0, 1}]
```

Question... Every now and then the imaginary number I shows up as a result of integration. But I'm doing a real integral. Why does this happen?

Answer... It is a lot easier (theoretically, anyway) for *Mathematica* to find antiderivatives using complex numbers. For example, $\int \frac{dx}{\sin x - \cos x}$ gives a complex answer:

```
Integrate[ 1/(Sin[x]-Cos[x]), x]
```

$$(1 + I) \, (-1)^{1/4} \, \text{ArcTan}\left[\left(\frac{1}{2} + \frac{I}{2}\right) (-1)^{1/4} \, \text{Sec}\left[\frac{x}{2}\right] \left(\text{Cos}\left[\frac{x}{2}\right] + \text{Sin}\left[\frac{x}{2}\right]\right)\right]$$

Nevertheless, you can check that the result is correct by differentiating it:

```
Simplify[D[%,x]]
```

$$\frac{1}{-\text{Cos}[x] + \text{Sin}[x]}$$

CHAPTER 13
Series, Taylor Series, and Fourier Series

Series

The Sum Command

You can use the **Sum** command to add up a finite number of terms of an indexed expression. To find $\sum_{n=n_0}^{n_1} expression$, you type:

> **Sum[** *expression* **, { n ,** n_0 **,** n_1 **}]**

For example,

> **Sum[n^2, {n, 1, 20}]**   (*Computes $\sum_{n=1}^{20} n^2$.*)
>
> 2870

> **Sum[Sin[n]/n, {n, 1, 5}]**
>
> $Sin[1] + \dfrac{Sin[2]}{2} + \dfrac{Sin[3]}{3} + \dfrac{Sin[4]}{4} + \dfrac{Sin[5]}{5}$

The **Sum** command has moderate success even with symbolic summations. For example, $\sum_{n=0}^{k} r^n = \frac{1-r^{k+1}}{1-r}$ is a partial sum for a geometric series:

> **Sum[r^n,{n,0,k}]**
>
> $\dfrac{-1 + r^{1+k}}{-1 + r}$

Numerical Summation and NSum

If you want an approximate value for a summation, use **NSum** instead. It has the same syntax as **Sum**.

> **NSum[** *expression* **, { n ,** n_0 **,** n_1 **}]**

For example,

> **NSum[1/n^2, {n, 1, 20}]**
>
> 1.59616

> **NSum[1/n^2, {n, 1, 10000}]**
>
> 1.64483

Infinite Series

NSum can also be used to find an approximate value of an infinite series. For example, $\sum_{n=1}^{\infty} \frac{1}{n^2}$ is known to converge to the value $\pi^2 / 6 \approx 1.64493$:

```
NSum[ 1/n^2, {n, 1, Infinity}]
1.64493
```

On the other hand, the harmonic series $\sum_{n=1}^{\infty} \frac{1}{n}$ diverges:

```
NSum[ 1/n, {n, 1, Infinity}]
NIntegrate::slwcon:
    Numerical integration converging too slowly;
    NIntegrate::ncvb: NIntegrate failed to converge
    to prescribed accuracy after 7 recursive
    bisections in n near n = 2.28833 10^56
23953.7
```

Do not think that the series converges to 23953.7! After reading the error messages carefully, you should suspect that the series diverges.

Some Famous Infinite Series

■ **Example.** Here are some well-known infinite series:

$$\sum_{n=0}^{\infty} \frac{1}{n!} = e, \qquad \sum_{n=0}^{\infty} \frac{1}{(2n+1)^2} = \frac{\pi^2}{8}, \quad \text{and} \quad \sum_{k=1}^{\infty} \frac{k}{e^{2\pi k} - 1} = \frac{1}{24} - \frac{1}{8\pi}$$

Mathematica version 3.0 recognizes the first two symbolically but can only compute the third numerically. (*Mathematica* 2.x cannot do the first two.)

```
Sum[ 1/n!, {n,0, Infinity}]
E

Sum[ 1/(2*n+1)^2, {n,0, Infinity}]
 π²
 ──
 8

NSum[k/(E^(2*Pi*k) -1), {k,1, Infinity}]
0.00187793

N[1/24 - 1/(8Pi)]
0.00187793
```

Taylor Series

Taylor Polynomials

Recall that if a function f satisfies certain reasonable conditions, then it can be approximated by a polynomial $p_n(x)$ of degree n near a point $x = a$ defined by:

$$p_n(x) = f(a) + \frac{f'(a)}{1!}(x - a) + \frac{f''(a)}{2!}(x - a)^2 + \cdots + \frac{f^{(n)}(a)}{n!}(x - a)^n$$

The polynomial $p_n(x)$ is called the Taylor polynomial of f of degree n about $x = a$.

We can use the **Sum** command to write out Taylor polynomials explicitly. For example, e^x has the following eighth-degree Taylor polynomial about $x = 0$:

```
f[x_] := Exp[x]
Sum[ (D[f[x],{x,k}] /. x->0) * x^k/k!, {k,0,8} ]
```

$$1 + x + \frac{x^2}{2} + \frac{x^3}{6} + \frac{x^4}{24} + \frac{x^5}{120} + \frac{x^6}{720} + \frac{x^7}{5040} + \frac{x^8}{40320}$$

Normal and Series Commands

A composition of the **Normal** and **Series** commands as follows will also produce the Taylor polynomial of degree n about $x = a$:

```
Normal[Series[ function, { x, a, n }] ]
```

For example,

```
Normal[Series[ Cos[x], {x, 0, 3}]]
```

$$1 - \frac{x^2}{2}$$

```
Normal[Series[ Cos[x], {x, 0, 8}]]
```

$$1 - \frac{x^2}{2} + \frac{x^4}{24} - \frac{x^6}{720} + \frac{x^8}{40320}$$

```
Normal[Series[ Cos[x], {x, 0.5, 8}]]
```

$0.877583 - 0.479426\,(-0.5 + x) - 0.438791\,(-0.5 + x)^2 +$
$0.0799043\,(-0.5 + x)^3 + 0.0365659\,(-0.5 + x)^4 -$
$0.00399521\,(-0.5 + x)^5 - 0.00121886\,(-0.5 + x)^6 +$
$0.0000951241\,(-0.5 + x)^7 + 0.0000217654\,(-0.5 + x)^8$

The **Series** command by itself actually gives a Taylor polynomial together with a remainder term, $O[x]$.

```
Series[ Cos[x], {x, 0, 8}]
```

$$1 - \frac{x^2}{2} + \frac{x^4}{24} - \frac{x^6}{720} + \frac{x^8}{40320} + O[x]^9$$

Applying **Normal** to this result will drop off the remainder term and give the Taylor polynomial.

```
Normal[%]
```

$$1 - \frac{x^2}{2} + \frac{x^4}{24} - \frac{x^6}{720} + \frac{x^8}{40320}$$

Fourier Series

FourierTrig-Series Command

You can ask *Mathematica* to find the Fourier Series of order n for a function in a given interval $a \le x \le b$. The command is:

```
Needs["Calculus`FourierTransform`"]
FourierTrigSeries[ function, { x , a , b }, n ]
```

For example, we can find the third-order Fourier Series for x^2 in the interval $0 \le x \le 1$ with

```
Needs["Calculus`FourierTransform`"]
FourierTrigSeries[ x^2, {x,0,1}, 3]
```

$$\frac{1}{3} + \frac{\text{Cos}[2\pi x]}{\pi^2} + \frac{\text{Cos}[4\pi x]}{4\pi^2} + \frac{\text{Cos}[6\pi x]}{9\pi^2} - \frac{\text{Sin}[2\pi x]}{\pi} - \frac{\text{Sin}[4\pi x]}{2\pi} - \frac{\text{Sin}[6\pi x]}{3\pi}$$

More Examples

Compare Graphically a Function with Its Taylor Polynomials

■ **Example.** Consider the function $f(x) = e^x + \sin(x)$:

```
f[x_] := Exp[x] + Sin[x]
pict1 =  Plot[f[x], {x,-3,3}, PlotStyle ->
                   {{Thickness[0.02], GrayLevel[0.5]}}]
```

Let us compare f graphically with some of its Taylor polynomials, say near $x = 1$. We start with the Taylor polynomial of degree 3 (but we will not show you the picture right away).

```
pict3 = Plot[
   Evaluate[Normal[Series[f[x], {x,1,3}]]], {x,-3,3}]
```

Notice that we used **Evaluate** with **Plot** above. This makes *Mathematica* compute the Taylor polynomial first before plotting the picture and is required because of the way that the **Plot** command works.

```
Show[pict1, pict3]
```
(*The graph of *f* is plotted in thick gray, and the Taylor polynomial is plotted in black.*)

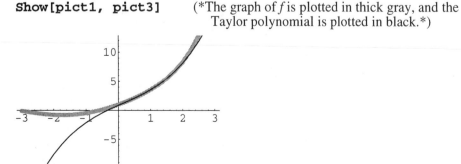

Now we increase the degree of the Taylor polynomial to 5 and then 9:

```
pict5 = Plot[
    Evaluate[Normal[Series[f[x], {x,1,5}]]], {x,-3,3}]

Show[pict1, pict5]
```

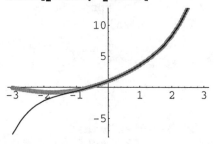

```
pict9 = Plot[
    Evaluate[Normal[Series[f[x], {x,1,9}]]], {x,-3,3}]

Show[pict1, pict9]
```

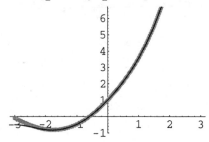

We can see that the approximation improves as we increase the degree of the Taylor polynomial.

Compare Graphically a Function with Its Fourier Series

■ **Example.** Consider the following function of split definition:

$$f(x) = \begin{cases} 0, & \text{if } -\pi \le x < 0 \\ 1, & \text{if } 0 \le x \le \pi \end{cases}.$$

Let us compare the graphs of f and its Fourier Series of orders 3, 5, and 9.

```
f[x_] := Which[ -Pi<=x<0, 0 , 0<=x<=Pi, 1 ]
```

(See Chapter 3 about the **Which** command.)

```
pict1 =  Plot[f[x], {x,-Pi,Pi},
                  PlotStyle -> {{Thickness[0.015]}}]

Needs["Calculus`FourierTransform`"]

p3[x_] = N[FourierTrigSeries[f[x],{x,-Pi,Pi},3]];
p5[x_] = N[FourierTrigSeries[f[x],{x,-Pi,Pi},5]];
p9[x_] = N[FourierTrigSeries[f[x],{x,-Pi,Pi},9]];
```

(We use "=" and "**N**[ ]" in the input lines above to force a numerical computation of the Fourier series.)

```
pict2 = Plot[{p3[x],p5[x],p9[x]},{x,-Pi,Pi},
                PlotStyle -> {{GrayLevel[0.8]},
                {GrayLevel[0.5]},{GrayLevel[0.2]}} ]
```

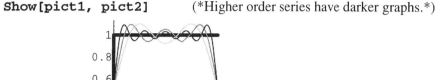

Show[pict1, pict2]   (*Higher order series have darker graphs.*)

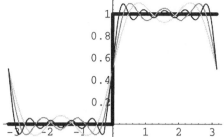

Troubleshooting Q & A

Question...   When I used **NSum** I got an error message that "`. . . is not numerical at point.`" What went wrong?

Answer...   Check whether you made an error in the input. Also check whether you used more than one variable in the expression. For example, if the variable **a** is unassigned, we can do

```
Sum[ n^2 + a^2, {n, 1, 10}]
```

but will get an error message with

```
NSum[ n^2 + a^2, {n, 1, 10}]
```

Question...   I had lots of trouble defining the Taylor Series or Fourier Series of a function. What should I check for?

Answer...   You should not use ":=" to define a Taylor Series or Fourier Series. You should use "=" instead. For example,

```
p[x_] = Normal[Series[ Cos[x], {x, 0.5, 8}]]
```

```
q[x_] = FourierTrigSeries[ x^2, {x,-1,1}, 3]
```

The reason is similar to what we explained for defining derivatives of functions in Chapter 11:

```
g[x_] := D[x^2, x]      (*This will not work.*)
g[x_] =  D[x^2, x]      (*This will work.*)
```

Also, the computation of the coefficients of a Fourier Series involves integrations that may be difficult for *Mathematica*. You can use **N** to ask for numerical approximations for the coefficients:

```
q[x_] = N[FourierTrigSeries[f[x],{x,-1,1},3]]
```

Question...   I ran into trouble when I tried to **Plot** a Taylor Series or Fourier Series. Why doesn't **Plot** work for me?

Answer...   You have to ask *Mathematica* to compute the Taylor or Fourier Series first, before plotting its picture. Thus, the command

```
Plot[Normal[Series[f[x],  {x,1,9}]],{x,-3,3}]
```

will not work. To make sure that *Mathematica* computes the Taylor Series first, you can either type

```
Plot[Evaluate[Normal[Series[f[x],{x,1,9}]]],{x,-3,3}]
```

or preferably

```
Clear[p,x]
p[x_] = Normal[Series[f[x],  {x,1,9}]]
Plot[ p[x],  {x,  -3,  3}]
```

CHAPTER 14
Solving Differential Equations

Symbolic Solutions of Equations

The DSolve Command

Using the **DSolve** command, you can symbolically solve ordinary differential equations that involve $y(x)$ and x. This command is used in the form:

> **DSolve[** *the differential equation* **, y[x] , x]**

> **Note:** The differential equation is entered using the double-equal sign **==** and $y(x)$ must appear as **y[x]**.

Here are some simple examples.

Equation	To solve it in Mathematica	Comment
$y' = 3x^2 y$	`DSolve[y'[x] == 3x^2*y[x], y[x], x]` $\left\{\left\{y[x] \to E^{x^3} C[1]\right\}\right\}$	y'[x] is used to denote y' in *Mathematica*.
$yy' = -x$	`DSolve[y[x]*y'[x] == -x, y[x], x]` $\left\{\left\{y[x] \to -\sqrt{-x^2 + 2\,C[1]}\,\right\},\right.$ $\left.\left\{y[x] \to \sqrt{-x^2 + 2\,C[1]}\,\right\}\right\}$	There are two solutions for this equation.
$y'' + y' - y = 0$	`DSolve[y''[x]+y'[x]-y[x]==0,y[x],x]` $\left\{\left\{y[x] \to E^{\frac{1}{2}\left(-1-\sqrt{5}\right)x} C[1] +\right.\right.$ $\left.\left. E^{\frac{1}{2}\left(-1+\sqrt{5}\right)x} C[2]\right\}\right\}$	This is a second-order equation. y''[x] denotes y''.

$y' = \cos(xy)$	`DSolve[ y'[x] == Cos[x*y[x]],`
	 `         y[x], x]`
	
	 `DSolve[ y'[x] == Cos[x*y[x]],y[x],x]`
		Mathematica returns the **DSolve** command unevaluated if it cannot solve the differential equation.

> **Note:** C[1] and C[2] denote arbitrary constants in the solutions above.

Equations with Initial or Boundary Conditions

Sometimes you need to solve a differential equation subject to initial or boundary conditions. In such a case, the format of the **DSolve** command is:

> **DSolve[{** *a differential equation* **,** *initial or boundary condition(s)* **},**
> **y[x], x]**

Here are some examples:

Equation & Condition(s)	To Solve It in Mathematica
$y' = -36x$, with $y(0) = 2$	`DSolve[{ y'[x] == -36x, y[0] == 2}, y[x], x]` $\{\{y[x] \to 2 - 18\,x^2\}\}$
$y' = 3x^2 y$ with $y(1) = -1$	`DSolve[{y'[x] == 3x^2*y[x], y[1] == -1}, y[x], x]` $\{\{y[x] \to -E^{-1+x^3}\}\}$
$y'' + y = 20\cos(x)$ with $y(0) = 0$, $y(\pi/2) = 1$	`DSolve[{y''[x]+y[x] == 20Cos[x], y[0]==0, y[Pi/2]==1}, y[x], x]` $\{\{y[x] \to -(-1 + 5\,\pi)\,\text{Sin}[x] + 10\,x\,\text{Sin}[x]\}\}$
$y'' - 2y' + y = x^2$ with $y(0) = 1$, $y'(0) = 2$	`DSolve[{y''[x]-2y'[x]+y[x] == x^2, y[0]==1, y'[0]==2}, y[x], x]  // Simplify` $\{\{y[x] \to 6 - 5\,E^x + (4 + 3\,E^x)\,x + x^2\}\}$

Notice that in the last two examples above, we needed two initial or boundary conditions to guarantee a unique solution for the given second-order equation.

Numerical Solutions of Equations

The NDSolve Command

The **NDSolve** command is used to find a numerical approximation to a solution of a differential equation over a specified interval $a \le x \le b$, subject to given initial conditions. It works even when **DSolve** fails. You use it in the following format:

> **NDSolve[{** *differential equation* **,** *initial condition(s)* **},**
> **y[x], {x, a , b }]**

For example, you can find a numerical approximation for the solution to the equation $y' = -xy$, subject to the initial condition $y(0) = 1$ over the interval $0 \le x \le 2$ with:

```
solution = NDSolve[{y'[x] == -x*y[x], y[0] == 1},
           y[x], {x,0,2}]
```

```
{{y[x] -> InterpolatingFunction[{0., 2.}, <>][x]}}
```

This output looks strange. Don't worry! *Mathematica* reports the answer as an `InterpolatingFunction`, which represents an algorithm to approximate the solution numerically.

To work with this `InterpolatingFunction`, we name it *f* using the following syntax. Note that we use = in defining the function. (This is similar to the way we extract solutions returned by the **Solve** command. See Chapter 5.)

```
Clear[f]
f[x_] = y[x] /. solution[[1]]
```

```
InterpolatingFunction[{0., 2.}, <>][x]
```

You can now compute the values $y(0)$, $y(0.25)$, ... , $y(1.75)$ and $y(2)$ with:

Table[f[x], {x, 0, 2, 0.25}]

```
{1., 0.969239, 0.882502, 0.754842, 0.606532,
   0.457836, 0.324654, 0.216267, 0.135335}
```

Or you can see a graph of the solution with:

Plot[f[x], {x, 0, 2}]

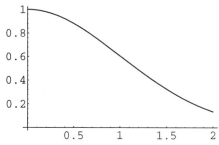

> **Note:** Make sure you use the syntax above to extract the result of **NDSolve**. You should not type the name **InterpolatingFunction** by hand.

Systems of Differential Equations

Solving a System Symbolically

The **DSolve** command can solve a system of differential equations involving two functions $x(t)$ and $y(t)$. You use it in the format:

DSolve[{ *system of differential equations* **}, {x[t], y[t]}, t]**

For example, to solve for functions $x(t)$ and $y(t)$ in the system $x'(t) = x(t) - y(t)$ and $y'(t) = y(t)$, use:

**DSolve[{x'[t] == x[t] - y[t], y'[t]== y[t]},
           {x[t], y[t]}, t]**

$$\left\{\left\{y[t] \to E^t\, C[2]\,,\, x[t] \to E^{E^t\, -\, C[2]}\, C[1]\right\}\right\}$$

> **Note:** You have to type **{ x[t], y[t] }** to denote the *two* functions in the command above.

Solving a System Numerically

NDSolve can be used to solve systems of differential equations numerically. For example, to solve the system $x'(t) = -2x(t)^2 - y(t)$ and $y'(t) = x(t) - y(t)$ numerically with initial conditions $x(0) = 0.2$ and $y(0) = 0.1$ over the interval $0 \le t \le 20$, use:

```
solution = NDSolve[{x'[t] == -2(x[t])^2 -y[t],
      y'[t] == x[t]-y[t],x[0] == 0.2, y[0] == 0.1},
      {x[t], y[t]}, {t, 0, 20}]
```

{{x[t] → InterpolatingFunction[{{0., 20.}}, <>][t],
     y[t] → InterpolatingFunction[{{0., 20.}}, <>][t]}}

Again, the outputs from **NDSolve** are given by InterpolatingFunctions. To use them, you have to assign names to them such as *f* and *g*. Use this syntax:

```
f[t_] = x[t] /. solution[[1]]
```

InterpolatingFunction[{{0., 20.}}, <>][t]

```
g[t_] = y[t] /. solution[[1]]
```

InterpolatingFunction[{{0., 20.}}, <>][t]

Now, for example, you can find the values of $(x(0), y(0))$ and $(x(5), y(5))$ with:

```
{{f[0],g[0]}, {f[5],g[5]}}
```

{{0.2, 0.1}, {-0.00373084, -0.0154398}}

You can also see the solution curve $(x(t), y(t))$ (called a phase diagram):

```
ParametricPlot[ {f[t], g[t]}, {t, 0, 20},
                    PlotRange -> All]
```

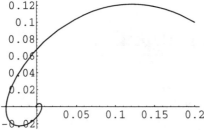

More Examples

Geometry of a First-Order Differential Equation

■ **Example.** Consider the differential equation $y' = 3x^2 y$. Let's find the general solution:

```
DSolve[ y'[x] == 3x^2*y[x], y[x], x]
```

$$\{\{y[x] \to E^{x^3} C[1]\}\}$$

You can sketch a few particular solutions of this equation, for example, with C[1] having values 0.5, –0.71 and 0.92. (We won't show the output here – you'll see the picture later anyway.)

```
solutions = Plot[{0.5 * Exp[x^3], -.71*Exp[x^3],
                    0.92*Exp[x^3]}, {x,-2,1},
                    PlotStyle -> Thickness[0.02]]
```

The slope field of this differential equation can be seen with the following command.

(We will explain this command in Chapter 20 when we discuss vector fields.)

```
Needs["Graphics`PlotField`"]
pict1 = PlotVectorField[{1, 3x^2*y},
              {x,-2,2}, {y,-2,2},
              ScaleFunction -> (1&), Frame -> True]
```

Now you see the solutions together with the slope field:

```
Show[pict1, solutions]
```

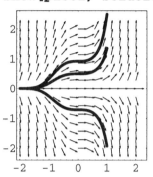

As you see from the picture, the solution curves are tangent to the vectors that make up the slope field. You may wonder why.

Recall that y' gives the slope of the graph of y at any point. Since each of the solution curves satisfies $y' = 3x^2y$, the tangent slope at each point will be $3x^2y$. At the same time, each of the vectors $(1, 3x^2y)$ in the slope field has slope $3x^2y$. This demonstrates that the solution curves are tangent to the slope field, as seen in the picture.

Troubleshooting Q & A

Question... I got an error message "not a valid equation or list of valid equations" for **DSolve** or **NDSolve**. What should I look for?

Answer... This most likely means that you did not enter the differential equation(s) correctly. The three most common mistakes made are these:

- You used = instead of == to enter the equation.
- You typed **y** in the equation instead of **y[x]**.
- You assigned values earlier to either the independent variable (**x**) or the function name (**y**). Try **Clear[x,y]**.

Question... I made the mistake of using an equal sign instead of the double-equal sign to define my differential equation. I've cleared the names **x** and **y** and evaluated **DSolve** again, but I'm still getting errors. What should I do?

Answer... Accidentally typing, say **DSolve[y'[x] = 3, y[x], x]** assigns the value 3 to the symbol **y'[x]**. Now the definition of the function **y** will be confused.

To fix the problem, you should use **Remove[x, y]** before retrying the **DSolve** command. (Note: **Clear** doesn't always do the job!) This eliminates any possible confusion about what **y'[x]** means.

Question...   I got **DSolve** to work without any error messages, but the solution it gave me doesn't seem to be right. Why?

Answer...   Make sure you used **y[x]** in the differential equation and not just **y** by itself. For example, **DSolve[y'[x] == y, y[x], x]** treats the **y** in the equation as a constant (not a function), and so it produces the solution $y(x) = x\,y + C$, rather than the expected solution $y(x) = Ce^x$.

Question...   When I used **NDSolve,** I got an error message "Insufficient initial conditions." What does this mean, and what should I do?

Answer...   Make sure you give the right number of initial conditions in your input. A first-order differential equation needs one initial condition, a second-order differential equation needs two initial conditions, and so on.

CHAPTER 15

Making Graphs in Space

Graphing Functions of Two Variables

Plot3D Command

The easiest way to sketch a surface in three dimensions (in "3-D") is to use the **Plot3D** command.

You input an expression that gives the height of a surface above the xy-plane, in terms of the independent variables x and y. You must also specify intervals $x_0 \le x \le x_1$ and $y_0 \le y \le y_1$. The **Plot3D** command then has the form:

> **Plot3D[** *an expression of x and y*, **{x,** x_0**,** x_1**}, {y,** y_0**,** y_1**}]**

For example, the surface whose height is $z = 4 - x^2 - y^2$ above the xy-plane, over the rectangle $-2 \le x \le 2$ and $-2 \le y \le 2$, is seen with:

Plot3D[4-x^2-y^2, {x, -2, 2}, {y, -2, 2}]

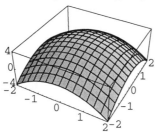

Mathematically speaking, this generates the surface $z = 4 - x^2 - y^2$ for $-2 \le x \le 2$ and $-2 \le y \le 2$. In the language of multivariable calculus, this means that **Plot3D** will show the graph of a two-variable function.

■ **Example.** The graph of the function $f(x,y) = x^2 - y^2$ looks like a saddle.

```
f[x_,y_] := x^2 - y^2
Plot3D[ f[x,y], {x, -3, 3}, {y, -3, 3} ]
```

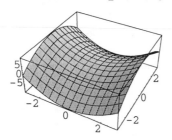

PlotPoints Option

You can use the **PlotPoints** option to generate a more detailed picture. Consider:

```
Plot3D[Sin[x*y], {x, -Pi, Pi}, {y, -Pi, Pi} ]
```

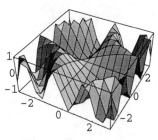

The picture does not look very good. But if we add the option **PlotPoints -> 30**, we'll see the graph shown with values sampled from a 30 × 30 grid, instead of the default 15 × 15 grid. This gives a smoother picture.

```
Plot3D[Sin[x*y], {x, -Pi, Pi}, {y,-Pi,Pi},
       PlotPoints -> 30]
```

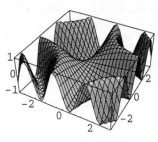

Other Useful Options

There are a number of options that you can use to add more "character" to a picture with **Plot3D**. For example,

Option	What It Does
AxesLabel -> {"x","y","z"}	Provides names to label each of the three axes.
BoxRatios -> {1, 1, 1}	Scales the graphic so that it appears within a cubical box.
BoxRatios -> Automatic	Scales the graphic so that units in each direction have the same length.
PlotRange -> All	Shows the whole picture.
Boxed -> False	No bounding box is drawn.
Axes -> None	Axis units are not labeled.

■ **Example.** The sombrero has equation $f(x,y) = \sin(\sqrt{x^2 + y^2})/\sqrt{x^2 + y^2}$:

```
Plot3D[ Sin[Sqrt[x^2+y^2]]/Sqrt[x^2+y^2],
    {x, -7,7}, {y,-7,7}, PlotPoints -> 25,
    BoxRatios -> {1,1,1},
    AxesLabel -> {"x","y","z"},
    PlotRange -> All]
```

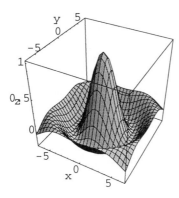

ViewPoints for 3-D Graphics

Every three-dimensional picture drawn in *Mathematica* is shown from a **ViewPoint**. Specifying a **ViewPoint** is the same as describing how your eye is situated with respect to the graphic being drawn. Let's show you how it's done.

The ViewPoint Option

You input the **ViewPoint** option in the form **ViewPoint –> { a, b, c }**, where the values a, b, and c describe the position of your eye relative to the object that you're viewing.

Positive values place you "in front of," "to the right of," and "above" the object, with respect to the direction of the positive x-, y- and z-coordinate axes, respectively. Negative values place you "behind," "to the left of," and "below" the object.

■ **Example.** The twin mountains:

```
f[x_,y_] := Exp[(-x^2/400)-(y^2/100)]*(x^2+y^2)

Plot3D[ f[x,y], {x, -40,40},{y,-40,40},
          AxesLabel -> {"x","y","z"}]
```

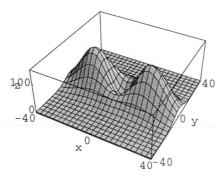

Now, let us take a "helicopter ride" and look at these mountains from different **ViewPoint**s.

```
Plot3D[ f[x,y], {x, -40,40},{y,-40,40},
    AxesLabel -> {"x","y","z"}, ViewPoint ->{3,1,1}]
```

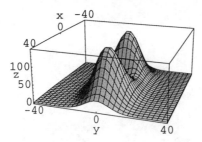

Show[%, ViewPoint -> {3, 2,-1}]   (*Look at it from the bottom.*)

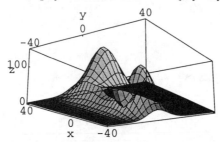

Show[%, ViewPoint -> {0,100,0}]   (*Look at it from the side.*)

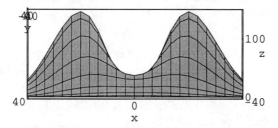

Note: *Mathematica*'s default **ViewPoint** is different from the standard position of 3-D graphics used in most Calculus books. (See the pictures below.) To achieve the standard view point, you have to add **ViewPoint –> { 3, 1, 1 }**.

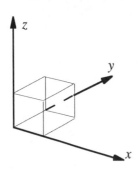

Mathematica's default
view point

Standard view point used
in most mathematics books

Surfaces in Cylindrical and Spherical Coordinates

Sometimes a surface is described in terms of the cylindrical or spherical coordinate systems. *Mathematica* can draw such a surface easily with the **CylindricalPlot3D** and **SphericalPlot3D** commands, respectively. These are defined in the **Graphics`ParametricPlot3D`** package.

**Cylindrical-
Plot3D
Command**

Points in the cylindrical coordinate system are described by quantities r, θ, and z, where

- r is the horizontal radial distance of the point from the z-axis;

- θ is the horizontal angle measured from the x-axis; and

- z is the z-coordinate in standard rectangular coordinates.

To draw the surface $z = f(r,\theta)$ for $r_0 \leq r \leq r_1$ and $\theta_0 \leq \theta \leq \theta_1$, you enter:

```
Needs["Graphics`ParametricPlot3D`"];
CylindricalPlot3D[ f(r,θ), {r, r₀, r₁}, {theta, θ₀, θ₁}]
```

> **Note:** In the **CylindricalPlot3D** command, you must enter the interval for **r** first, then the interval for **theta**. If you do not follow this order, the picture will be incorrect.

For example, to see the surface $z = r^2 \cos(5\theta)$, for $0.5 \leq r \leq 1$ and $0 \leq \theta \leq 2\pi$:

```
Needs["Graphics`ParametricPlot3D`"];
CylindricalPlot3D[ r^2*Cos[5*theta], {r,0.5,1},
                   {theta,0, 2Pi}]
```

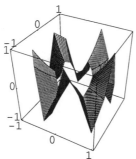

Notice that this ribbon-like surface is very "choppy" as the horizontal angle θ varies, yet the surface is quite smooth along the r–radial direction. This suggests that we should increase the resolution of the graphic in the θ variable (from 15 to 60) but decrease the resolution in r (from 15 to 10). We can do this with:

```
CylindricalPlot3D[ r^2*Cos[5*theta], {r,0.5,1},
                   {theta,0, 2Pi}, PlotPoints -> {10,60}]
```

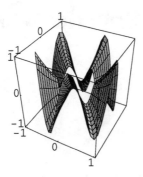

Spherical-Plot3D Command

Points in the spherical coordinate system are described by quantities ρ, θ, and ϕ, where

- ρ is the radial distance in space of the point from the origin;
- θ is the horizontal angle measured from the x-axis; and
- ϕ is the vertical angle measured from the z-axis.

To draw the surface $\rho = f(\theta, \phi)$, $\theta_0 \le \theta \le \theta_1$, and $\phi_0 \le \phi \le \phi_1$, you will enter:

```
Needs["Graphics`ParametricPlot3D`"];
SphericalPlot3D[ f(θ, φ), {theta, θ₀, θ₁}, {phi, φ₀, φ₁}]
```

> **Note:** In the **SphericalPlot3D** command, you have to enter the interval for **theta** first, then the interval for **phi**. If you do not follow this order, the picture will be incorrect.

For example, to see the surface $\rho = \sqrt{\theta}\,(3 + \cos\phi)$, $0 \le \theta \le 3\pi/2$, and $0 \le \phi \le \pi$:

```
Needs["Graphics`ParametricPlot3D`"];
SphericalPlot3D[ Sqrt[theta]*(3 + Cos[phi]),
      {theta,0,3 Pi/2}, {phi, 0, Pi}]
```

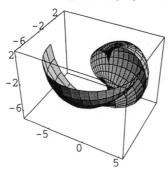

More Examples

Plot3D vs. Cylindrical-Plot3D

■ **Example.** In some cases, a surface given in rectangular coordinates will look better if you draw it using cylindrical or spherical coordinates. For example, the surface $z = \frac{x^2 - y^2}{(x^2 + y^2)^2}$ can be plotted with:

```
Plot3D[ (x^2 - y^2)/(x^2+y^2)^2, {x,-3,3}, {y,-3,3}]
```

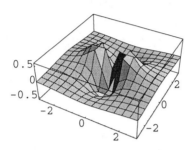

The picture is choppy especially near the origin. However, if we use cylindrical coordinates, the surface becomes

$$z = \frac{x^2 - y^2}{(x^2 + y^2)^2} = \frac{(r\cos\theta)^2 - (r\sin\theta)^2}{((r\cos\theta)^2 + (r\sin\theta)^2)^2} = \frac{r^2(\cos^2\theta - \sin^2\theta)}{r^4} = \frac{\cos 2\theta}{r^2}.$$

```
CylindricalPlot3D[ Cos[2*theta]/r^2, {r, 0.05, 3},
          {theta,0,2*Pi}, ViewPoint -> {3,2,1}]
```

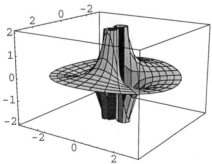

This picture is nicer!

Helpful Tips

💡 💡 Usually, you should not worry about specifying a **ViewPoint** the first time you sketch a 3-D graphic. After you've seen the graphic, you can redraw it from a different view point, using:

```
Show[ %, ViewPoint -> { a , b , c }]
```

Troubleshooting Q & A

Question... *Mathematica* either quit running or "crashed" when I was drawing some 3-D graphics. I also saw some messages that *Mathematica* "... needs more memory". What should I do?

Answer... *Mathematica* uses a lot of memory when it creates 3-D pictures. You need a very capable system with lots of memory to run *Mathematica*. Most "crashes" occur because there's not enough memory on your system.

When you installed *Mathematica* "out of the box," your system memory parameters may not have been set high enough to sustain intensive 3-D imaging. If you can increase *Mathematica*'s memory partition on your system, please do so.

Question... I tried to draw a 3-D picture but got an error message telling me that "Options expected ... beyond position 2." What happened?

Answer... A common mistake is to try to use the **Plot** command to draw a 3-D picture. Make sure you use the **Plot3D** command. For example, in the command:

      Plot[x^2+y^2, {x,-2,2}, {y,-2,2}]

Mathematica interprets {**y, –2, 2**} to be an option for **Plot**!

Question... When I used **Plot3D**, **CylindricalPlot3D** or **SphericalPlot3D**, I got an error message that the expression is not a "machine-size real number" or not a "real number." What should I check for?

Answer... This means that *Mathematica* cannot evaluate your input function numerically. Check whether:

- You mistyped the input.
- You used the same variables in the function as you used in specifying the intervals.
- The function is well-defined in the given intervals.

Question... I drew the graph of a function using **Plot3D**, but the picture looked different from the one shown in my calculus book. Why is that?

Answer... Remember that *Mathematica*'s 3-D output is not viewed from the standard position used by most math books. We recommend that you add the option **ViewPoint –>** { **3, 1, 1** } to correct it.

Also, two graphics may sometimes look different because they are drawn with different scales on the three axes. Adjust them with the **BoxRatios** option.

Question... When I used **CylindricalPlot3D** or **SphericalPlot3D**, nothing happened. What should I do?

Answer... Did you remember to load the package **Graphics`ParametricPlot3D`**? If you didn't, then here is what you should type:

```
Remove[CylindricalPlot3D, SphericalPlot3D]
Needs["Graphics`ParametricPlot3D`"]
```

Now, try the command again.

Question...   How do I draw two graphs with a single **Plot3D** command?

Answer...   You cannot draw multiple surfaces with **Plot3D** in 3-D like you do with **Plot** in two dimensions. In 2-D you can use:

```
Plot[{x, x^2}, {x,-2,2}]
```

to draw the graphs $y = x$, $y = x^2$ together. But **Plot3D** gives an error:

```
Plot3D[{x^2+y^2, x+y}, {x,-2,2}, {y,-2,2}]
```
Plot3D::plnc : $\{x^2 + y^2, x+y\}$ is neither a machine-

size real number at {x, y}={-2., -2.} nor a list
of a real number and a valid color directive.

You must draw the pictures separately and then use **Show** command to combine the two pictures:

```
pict1 = Plot3D[ x^+y^2, {x,-2,2}, {y,-2,2}]
pict2 = Plot3D[ x+y, {x,-2,2}, {y,-2,2}]

Show[ pict1, pict2]
```

Question...   The picture I got from **CylindricalPlot3D** or **SphericalPlot3D** was completely wrong. What should I check?

Answer...   Check these three areas:

- Make sure you typed the input function and the intervals of the two parameters correctly.

- In **CylindricalPlot3D** you have to enter the radius interval $\{\mathbf{r}, r_0, r_1\}$ first, followed by the angle interval $\{\mathbf{theta}, \theta_0, \theta_1\}$. If you enter these in the wrong order, *Mathematica* will reverse the sense of the variables.

- In **SphericalPlot3D** you have to enter the θ-interval $\{\mathbf{theta}, \theta_0, \theta_1\}$ first, then the ϕ-interval $\{\mathbf{phi}, \phi_0, \phi_1\}$. If you enter these in the wrong order, *Mathematica* will draw an incorrect picture.

CHAPTER 16
Level Curves and Level Surfaces

Level Curves in the Plane

The ContourPlot Command

In *Mathematica*, the level curves (contours) of a function $f(x, y)$ are plotted with the **ContourPlot** command. To see the level curves inside the rectangle $x_0 \leq x \leq x_1$, $y_0 \leq y \leq y_1$, you use the command:

```
ContourPlot[ function , {x, x₀, x₁}, {y, y₀, y₁}]
```

(The syntax looks exactly like the **Plot3D** command syntax that we discussed in the previous chapter.) For example, here are some level curves of $f(x, y) = x y e^{-x^2 - y^2}$:

```
ContourPlot[x*y*Exp[-x^2-y^2], {x,-2,2}, {y,-2,2}]
```

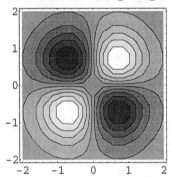

Mathematica shades the areas between level curves. Lighter shades represent higher levels, while darker shades represent lower levels. Thus, we can tell from this picture that the function has its largest values near $(1, 1)$ and $(-1, -1)$.

Options for ContourPlot

Some of the options we like to use with **ContourPlot** are the following:

Option	What it Does
ContourShading -> False	Shows the level curves without shading between them.
Contours -> n	Draws n level curves.
PlotPoints -> n	Increases the resolution of the picture.

Here's a nicer picture than the previous one:

```
ContourPlot[x*y*Exp[-x^2-y^2], {x,-2,2}, {y,-2,2},
      PlotPoints -> 100, Contours -> 20]
```

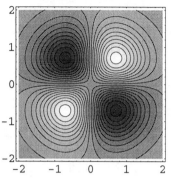

> **Note:** The default value for **PlotPoints** is 15. Be careful. Larger values can impact execution timing significantly.

Plotting Specific Levels

You can plot specific level curves using the **Contours** option. You use it in the form **Contours** –> { *the levels* }. The levels must be separated by commas.

■ **Example**. To see the contours of $f(x, y) = x y e^{-x^2-y^2}$ at levels 0, 0.1, and 0.15, without the shading:

```
ContourPlot[x*y*Exp[-x^2-y^2], {x,-2,2}, {y,-2,2},
            Contours -> {0, 0.1, 0.15},
            PlotPoints->100,
            ContourShading -> False]
```

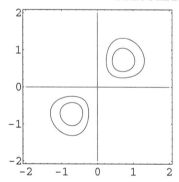

Notice that the level curve of $f(x, y) = x y e^{-x^2-y^2}$ at level 0 consists of the x- and y-axes. This is why the axes are shown in the picture.

Level Surfaces in Space

The ContourPlot3D Command

If f is a function of three variables defined over a region $x_0 \leq x \leq x_1$, $y_0 \leq y \leq y_1$, and $z_0 \leq z \leq z_1$, then the level surface of f at level c can be seen with the **ContourPlot3D** command. It's defined in one of the **Graphics** packages.

```
Needs["Graphics`ContourPlot3D`"]
ContourPlot3D[ function, {x, x₀, x₁}, {y, y₀, y₁},
                     {z, z₀, z₁}, Contours -> { c } ]
```

You can specify more than one level surface to be shown in the same graphic by writing **Contours** –> { *the levels* }, where the levels are separated by commas.

Here are the level surfaces of $f(x, y, z) = x^3 - y^2 + z^2$ at the levels of 1 and 10:

```
Needs["Graphics`ContourPlot3D`"]
ContourPlot3D[ x^3-y^2+z^2, {x,-2,5}, {y,-2,2},
                  {z,-2,3}, Contours -> {1,10}]
```

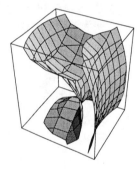

That is, the surfaces shown have equations $x^3 - y^2 + z^2 = 1$ and $x^3 - y^2 + z^2 = 10$.

> **Note: ContourPlot3D** produces very nice graphics but unfortunately requires significant computation time.

More Examples

Comparing Plot3D with ContourPlot

■ **Example**. Consider $f(x, y) = x^2 - y^2$. The following command shows the contours at levels 0, 1 and –1.

```
ContourPlot[x^2-y^2, {x,-2,2}, {y,-2,2},
                 Contours->{0, 1, -1}]
```

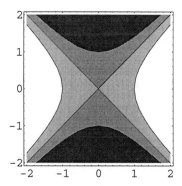

What are these curves? Recall that a contour of f at level c is given by the equation $f(x, y) = c$. We thus have:

- The contour at level 0 has equation $x^2 - y^2 = 0$, which consists of the two lines $y = x$ and $y = -x$.

- The contour at level 1 has equation $x^2 - y^2 = 1$ and is a hyperbola that opens left/right.

- The contour at level –1 has equation $x^2 - y^2 = -1$ and is also a hyperbola that opens up/down.

We can see how these three contours arise by intersecting the graph of $f(x, y) = x^2 - y^2$ with the planes $z = 0$, $z = 1$ and $z = -1$.

```
pict1 = Plot3D[ x^2-y^2, {x,-2,2}, {y,-2,2}]
pict2 = Plot3D[ 0,{x,-2,2},{y,-2,2},PlotPoints->2]
pict3 = Plot3D[ 1,{x,-2,2},{y,-2,2},PlotPoints->2]
pict4 = Plot3D[ -1,{x,-2,2},{y,-2,2},PlotPoints->2]
Show[pict1, pict2, pict3, pict4,
     BoxRatios -> {1,1,1}, ViewPoint -> {1,3,0.7}]
```

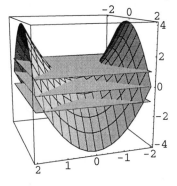

Curves in the Plane Defined by an Equation

■ **Example**. The equation $2x^2 - 3xy + 5y^2 - 6x + 7y = 8$ defines a rotated ellipse in the plane. We could use **ImplicitPlot** to draw it. But it's also just the level curve of the function $f(x, y) = 2x^2 - 3xy + 5y^2 - 6x + 7y$ at level 8. We can see it with:

```
ContourPlot[2x^2 - 3x*y + 5y^2 - 6x + 7y,
      {x,-2,5}, {y,-3,2}, Contours -> {8},
      ContourShading -> False, PlotPoints -> 50]
```

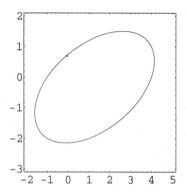

(The intervals used in the command above were arrived at after some experimentation, so that we could give you a nice picture.)

The Quadric Surfaces in Space

The Quadric Surfaces are those surfaces in space which can be given by an equation of the form:

$$Ax^2 + By^2 + Cz^2 + Dxy + Eyz + Fxz + Gx + Hy + Iz + J = 0,$$

where $A, B, C, \ldots, J$ are constants. These surfaces are discussed in detail in every multivariable calculus book. With the help of the **ContourPlot3D** command, we can easily see pictures of various quadric surfaces.

■ **Example**. The equation $3x^2 + 4y^2 + 5z^2 = 9$ defines an ellipsoid. It is just the level surface of the function $f(x, y, z) = 3x^2 + 4y^2 + 5z^2$ at level 9.

```
Needs["Graphics`ContourPlot3D`"]
ContourPlot3D[3x^2 + 4y^2 + 5z^2 , {x,-2,2},
            {y,-2, 2}, {z,-2,2}, Contours->{9},
            ViewPoint -> {2,1,1}]
```

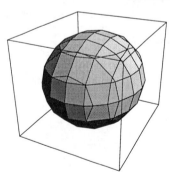

■ **Example**. The equation $\dfrac{x^2}{2^2} + \dfrac{y^2}{3^2} - \dfrac{z^2}{4^2} = 1$ defines a hyperboloid of one sheet:

```
Needs["Graphics`ContourPlot3D`"]
ContourPlot3D[x^2/2^2 + y^2/3^2 - z^2/4^2,
            {x,-10,10}, {y,-10,10}, {z,-10,10},
            Contours->{1}, ViewPoint->{2,1,1}]
```

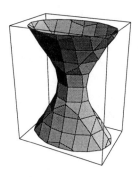

Troubleshooting Q & A

Question... I got an empty picture from **ContourPlot** when I specified a list of levels in the **Contours** option. What happened?

Answer... There are three likely possibilities:

- The level curves you specified do not exist, so no curves or points are shown in the output (e.g., the contour with level 0 is empty for $f(x, y) = x^2 + y^2 + 1$).

- The level curves you wanted to see do not lie inside the rectangle you gave (e.g., the contour $x^2 + y^2 = 1$ doesn't lie inside $2 \le x \le 3$, $2 \le y \le 3$).

- You input the function incorrectly (e.g., you typed **xy** instead of **x*y**, although you should have gotten some error messages which indicated this before you got the empty graphic).

Question... I got a black box from **ContourPlot** when I specified a list of levels in the **Contours** option. What happened?

Answer... Every point inside your region is at a level lower than the contours you specified. As a result, every point in the rectangle is darkly shaded. Readjust your list of contours.

Question... I tried to use **ContourPlot3D**, but *Mathematica* returned my input unevaluated. What happened?

Answer... You probably did not load the package which defines **ContourPlot3D** before you used it. Type the following, and then retry your **ContourPlot3D** command:

```
Remove[ContourPlot3D]
Needs["Graphics`ContourPlot3D`"]
```

Question... I got an empty picture from **ContourPlot3D**. Only the usual 3-D framebox appeared. What happened?

Answer... The level surface you're trying to see doesn't lie inside the region $x_0 \le x \le x_1$, $y_0 \le y \le y_1$, and $z_0 \le z \le z_1$ you gave. (This is similar to the problem addressed in the first question above.) Recheck both the level and the region you specified.

CHAPTER 17
Partial Differentiation and Multiple Integration

Partial Derivatives

The D Command

The **D** command we used in Chapter 11 is actually a partial differentiation operator. It differentiates an expression with respect to a specified variable, treating all other symbols as constants.

> **D[** *the function or expression* **,** *variable* **]**

For example, if $f(x, y) = 3xy^2 - 5y \sin x$, its partial derivatives $f_x = \dfrac{\partial f}{\partial x}$ and $f_y = \dfrac{\partial f}{\partial y}$ are computed with

```
f[x_,y_] := 3x*y^2 - 5y*Sin[x]
D[f[x,y], x]
```

$3 y^2 - 5 y \cos[x]$

```
D[f[x,y], y]
```

$6 x y - 5 \sin[x]$

You find higher-order derivatives by listing the variables in the order of differentiation. For example, $f_{xxy} = \dfrac{\partial^3 f}{\partial y \partial x^2}$ represents taking the partial derivative "first by x, then by x, then by y." You compute this with

```
D[f[x,y], x, x, y]
```

$5 \sin[x]$

Double and Triple Integrals

Iterated Double Integrals

Mathematica can do an iterated double integral with the **Integrate** command. The iterated double integral $\int_a^b \int_{g_1(x)}^{g_2(x)} f(x, y)\, dy\, dx$ is evaluated in the form:

> **Integrate[** *function* **,** **{x,** *a*, *b*}**,** **{y,** $g_1(x)$, $g_2(x)$}**]**

Similarly, the double iterated integral $\int_{c}^{d} \int_{h_1(y)}^{h_2(y)} f(x, y)\, dx\, dy$ is computed with:

Integrate[*function* **, {y,** c**,** d**}, {x,** $h_1(y)$**,** $h_2(y)$**}]**

> **Note:** The limits of the outer integral are written first. This is different from the way you compute the integrations by hand.

■ **Example.** The double integral $\int_{0}^{2} \int_{0}^{2x} (3x^2 + (y-2)^2)\, dy\, dx$ is computed with:

Integrate[3x^2 + (y-2)^2, {x, 0, 2}, {y, 0, 2x}]

$$\frac{88}{3}$$

The integration above takes place over the region D defined by the inequalities:

$$0 \le x \le 2 \quad \text{and} \quad 0 \le y \le 2x$$

(See the picture to the right.) D can also be described by the inequalities:

$$0 \le y \le 4 \quad \text{and} \quad y/2 \le x \le 2$$

It follows that the double integral $\int_{0}^{2} \int_{0}^{2x} (3x^2 + (y-2)^2)\, dy\, dx$

has the same value as $\int_{0}^{4} \int_{y/2}^{2} (3x^2 + (y-2)^2)\, dx\, dy$:

Integrate[3x^2 + (y-2)^2, {y, 0, 4}, {x, y/2, 2}]

$$\frac{88}{3}$$

Iterated Triple Integrals

An iterated triple integral $\int_{a}^{b} \int_{g_1(x)}^{g_2(x)} \int_{h_1(x, y)}^{h_2(x, y)} f(x, y, z)\, dz\, dy\, dx$ is evaluated using this extended form of the **Integrate** command:

Integrate[*function* **,{x,** a**,** b**},**
  {y, $g_1(x)$**,** $g_2(x)$**}, {z,** $h_1(x, y)$**,** $h_2(x, y)$**}]**

Other variations in the order of integration can be written with appropriate changes in the order of specifying the variables.

For example, to evaluate $\int_{-3}^{3} \int_{-\sqrt{9-x^2}}^{\sqrt{9-x^2}} \int_{x+y}^{3+y} z^2\, dz\, dy\, dx$, write:

Integrate[z^2, {x, -3, 3},
  {y, -Sqrt[9-x^2], Sqrt[9-x^2]}, {z, x+y, 3+y}]

$$\frac{567\,\pi}{4}$$

NIntegrate Command

If you want to find a numeric approximation for a double or triple integral, you should use **NIntegrate**. It has the same format as the **Integrate** command. For example, to find $\int_{-3}^{3}\int_{-\sqrt{9-x^2}}^{\sqrt{9-x^2}}\int_{x+y}^{3+y} z^2 \, dz \, dy \, dx$ numerically, write:

```
NIntegrate[z^2, {x,-3,3},
      {y, -Sqrt[9-x^2], Sqrt[9-x^2]}, {z,x+y,3+y}]
445.321
```

As you may expect, **NIntegrate** will give you an answer quickly in most cases, and it can be used even when **Integrate** fails.

More Examples

Critical Points and Hessian Test

■ **Example.** Suppose $f(x,y) = x^4 - 3x^2 - 2y^3 + 3y + 0.5xy$. We can find its critical points by solving the equations $f_x = 0$ and $f_y = 0$ simultaneously:

```
f[x_,y_] := x^4 - 3x^2 - 2y^3 + 3y + 0.5x*y
criticalpoints = NSolve[
    {D[f[x,y],x] == 0, D[f[x,y],y] == 0},{x, y}];
N[criticalpoints,4]
```

```
{{y → -0.7776, x → 1.256}, {y → -0.7036, x → -0.05877},
  {y → -0.6326, x → -1.197}, {y → 0.6291, x → -1.25},
  {y → 0.7106, x → 0.05936}, {y → 0.7741, x → 1.191}}
```

There are six critical points. We will define the discriminant

$$D = (f_{xx})(f_{yy}) - (f_{xy})^2$$

and evaluate it at each critical point. The Hessian Test that you learned in multivariable calculus says:

- If $D < 0$, the critical point is a saddle point.

- If $D > 0$, the critical point is a local maximum when f_{xx} is negative.

- If $D > 0$, the critical point is a local minimum when f_{xx} is positive.

We compute the discriminant D and f_{xx} at each of the critical points:

```
discriminant = D[f[x,y],x,x]*D[f[x,y],y,y] -
                  D[f[x,y],x,y]^2
```

$-0.25 - 12 \left(-6 + 12 \, x^2\right) y$

```
{discriminant , D[f[x,y],x,x]} /. criticalpoints
{{120.39, 12.9287}, {-50.5617, -5.95855},
  {84.8317, 11.2076}, {-96.5455, 12.7549},
  {50.5524, -5.95772}, {-102.667, 11.0251}}
```

The result above shows that the discriminant is positive at the first, third, and fifth of the critical points. Since f_{xx} is positive at the first and third, those critical points [(1.256, –0.778) and (–1.197, –0.633)] will be local minima for f. Also, the point (0.059, 0.711) will be a local maximum.

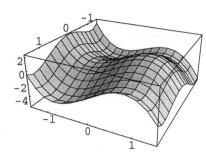

The three remaining critical points have negative discriminants, so each is a saddle point for f. You can check these points from the graph of f shown above.

Method of Lagrange Multipliers

■ **Example**. We want to find the maximum and minimum values of the function $f(x, y) = (x - 2)y + y^2$, subject to the constraint $x^2 + y^2 - 1 = 0$. The Method of Lagrange Multipliers states that we need to solve the system of equations:

$$f_x = \lambda g_x, \ f_y = \lambda g_y \text{ and } g = 0,$$

where g is the "constraint" function $g(x, y) = x^2 + y^2 - 1$.

```
f[x_,y_] := (x-2)*y + y^2
g[x_,y_] := x^2 + y^2 - 1

solutions = NSolve[ {D[f[x,y], x] == p* D[g[x,y],x],
   D[f[x,y], y] == p* D[g[x,y],y], g[x,y] == 0},
   {p, x, y}];
N[solutions,4]
```

```
{{p → -0.6484, x → -0.6107, y → 0.7919},
 {p → 0.2525 - 0.2668 I, x → 0.9189 + 0.238 I, y → 0.591 - 0.3701 I},
 {p → 0.2525 + 0.2668 I, x → 0.9189 - 0.238 I, y → 0.591 + 0.3701 I},
 {p → 2.143, x → - 0.2272, y → - 0.9739}}
```

(Note: We use the variable **p** in this computation to stand for the multiplier λ.) We can find the values of f at these points with:

```
f[x,y] /. solutions
```

```
{-1.44027, -0.338533 + 0.103341 I,
 -0.338533 - 0.103341 I, 3.11733}
```

So f has the minimum value –1.44027 at the first point (–0.61067, 0.79189) and the maximum value 3.11733 at the last point (–0.22717,–0.97386). The other answers are complex numbers so we can ignore them.

You can also see this result geometrically by drawing the contour picture of f and the constrained curve $g = 0$ together:

```
pict1 = ContourPlot[f[x,y], {x,-1.5,1.5},
   {y,-1.5,1.5}, Contours -> 30]
```

```
Needs["Graphics`ImplicitPlot`"]
pict2 = ImplicitPlot[ g[x,y] == 0, {x,-2,2},
   {y,-2,2}, PlotStyle -> {Thickness[0.02]}]
```

```
Show[pict1, pict2]
```

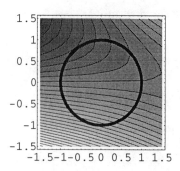

Recall that darker shades represent lower values of f in the graphic above. Thus the point $(-0.61067, 0.79189)$ on the constraint circle gives the minimum because that is where the darkest shading occurs.

Integration in Polar Coordinates

■ **Example.** Let us compute the integral $\iint_D e^{(x^2+y^2)} dA$, where D is the circular sector given in polar coordinates as $0 \le r \le 1$, $\pi/4 \le \theta \le \pi/2$. ($D$ is sketched to the right.)

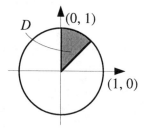

We learned from calculus that we must first write the integral as the iterated integral $\int_{\pi/4}^{\pi/2}\int_0^1 e^{r^2} r\, dr\, d\theta$. Now we can compute its value:

```
Integrate[ Exp[r^2]*r, {t, Pi/4, Pi/2}, {r, 0, 1}]
```

$$-\frac{\pi}{8} + \frac{E\,\pi}{8}$$

Outsmarting Mathematica

■ **Example.** You can outsmart *Mathematica* easily. Let D be the region inside the circle $x^2 + y^2 = 1$. Find the double integral $\iint_D \sqrt[3]{x^2 + y^2}\, dA$ both by hand and by computer.

- For the computer: Evaluate the double integral using rectangular coordinates. The region D is described by the inequalities $-1 \le x \le 1$ and $-\sqrt{1-x^2} \le y \le \sqrt{1-x^2}$, so you enter:

```
Integrate[ (x^2+y^2)^(1/3), {x,-1,1},
          {y, -Sqrt[1-x^2], Sqrt[1-x^2]} ]
```

- For you, while you're waiting for the computer to finish: Do the integration in a more "civilized" way. You can evaluate the integral directly using polar coordinates, because the region D is described by the inequalities $0 \le r \le 1$ and $0 \le \theta \le 2\pi$:

$$\iint_D \sqrt[3]{x^2 + y^2}\, dA = \int_0^{2\pi}\int_0^1 (r^2)^{1/3} r\, dr\, d\theta = \int_0^{2\pi}\int_0^1 (r^{5/3})\, dr\, d\theta$$

$$= 2\pi(\tfrac{3}{8} r^{8/3}\Big|_0^1) = \tfrac{3\pi}{4}.$$

Interrupt the computer's calculation if it's still running. (If the computer has "finished," figure out whether the answer it gave was sensible!)

You Win!!!

Troubleshooting Q & A

Question...   I entered a double (or triple) integral expression and have been waiting for a response for two or three minutes. *Mathematica* hasn't given me an answer yet. Is something wrong?

Answer...   Integration is a difficult mathematical problem. Unfortunately, *Mathematica* cannot solve every integration problem you come up with.

You may have to abort your calculation and look for ways to simplify the integrand (e.g., through use of cylindrical or spherical coordinates in three variables) in order to move toward getting a result.

It is also not uncommon for a slow computer to take several minutes to evaluate a complicated double or triple integral. The fact is that you may simply need a faster computer or more patience. Sad, but true!

Question...   *Mathematica* did not give me a number when I evaluated a double or triple integral. But I know the answer should be numerical. What should I look for?

Answer...   Check for these possible problem areas:

- It could be a mathematical error. Did you set up the integral correctly? Recheck your limits of integration, too.

- It could be a *Mathematica* input error. Did you type the integrand correctly? Did you input the order of integration correctly? (Remember that the outer limits of integration must be entered first – see the Note earlier in this chapter.)

Question...   When I used **NIntegrate** I got an error message which said "Integrand ... is not numerical at {x, y} =." What does this mean?

Answer...   The function you want to integrate is either typed incorrectly or has variables that are not involved in the integration and that don't evaluate as numbers. For example, your input function might be written in terms of *x* and *y*, but you're doing the integration with respect to polar coordinates **theta** and **r**.

Question...   When I used **NIntegrate** I got an error message that "... is not a valid limit of integration." What should I look for?

Answer...   Check that you input the order of integration correctly. (Remember that the outer limits of integration must be entered first – see the Note earlier in this chapter.)

Also check the limits of integration carefully. You may have mistyped them, or you may have used variables that are not involved in the integration.

CHAPTER 18

Matrices and Vectors

Vectors

Defining Vectors

You use lists to define vectors in *Mathematica*. For example, the vectors (2, 3) and (−1, 1, 2) are entered as

```
{2, 3}
```
$\{2, 3\}$

```
{-1, 1, 2}
```
$\{-1, 1, 2\}$

You do addition, subtraction, and scalar multiplication of vectors directly using the operations **+**, **−** and *****. *Mathematica* applies these quite naturally:

```
{1, 1, 2} + { 2, -1, 3}
```
$\{3, 0, 5\}$

```
{a1, b1, c1} - k*{a2, b2, c2}
```
$\{a1 - a2\,k,\ b1 - b2\,k,\ c1 - c2\,k\}$

Dot Product

You compute the dot product of two vectors using the **Dot** command:

```
Dot[ {a1, b1, c1}, {a2, b2, c2} ]
```
$a1\,a2 + b1\,b2 + c1\,c2$

Mathematica also allows a shorthand, in-line notation for **Dot** by typing a period "." (literally, a "dot"!):

```
{a1, b1, c1} . {a2, b2, c2}
```
$a1\,a2 + b1\,b2 + c1\,c2$

Cross Product

To compute a cross product of two 3-D vectors (also known as the vector product), you use the **Cross** command. This is built-in for version 3.0 of *Mathematica*, but for earlier versions, you must load the definition of the **Cross** command from one of the **LinearAlgebra** packages first:

```
Needs["LinearAlgebra`CrossProduct`"]    (*before Vers. 3.0*)
Cross[ {a1, b1, c1}, {a2, b2, c2}]
```
$\{-b2\,c1 + b1\,c2,\ a2\,c1 - a1\,c2,\ -a2\,b1 + a1\,b2\}$

There are many other useful commands defined in the **LinearAlgebra** collection of packages. You can load all of them with

Needs["LinearAlgebra`Master`"]

This package contains commands such as **GramSchmidt**, **SubMatrix**, **LUFactor**, and **LUSolve**, which are very helpful in studying matrix theory or linear algebra.

Matrices

Defining Matrices

You define a matrix in *Mathematica* by entering each row as a list. For example, to enter the matrix $\begin{pmatrix} 3 & -4 & 7 \\ -1 & 0 & 5 \end{pmatrix}$, you type:

{ {3,-4, 7}, {-1,0,5} }

${\{3, -4, 7\}, \{-1, 0, 5\}\}}$

Here, each row of the matrix is a list of three elements. The matrix is formed as a list of its two rows.

Mathematica really does think the entry above is a matrix. To see this use the **MatrixForm** command:

MatrixForm[{ {3,-4, 7}, {-1,0,5} }]

or use it in its postfix syntax:

{ {3,-4, 7}, {-1,0,5} } // MatrixForm

$\begin{pmatrix} 3 & -4 & 7 \\ -1 & 0 & 5 \end{pmatrix}$

Basic Operations involving Matrices

You can do addition, subtraction, scalar multiplication, and matrix multiplication with the operators "+", "−", "*", and ".", respectively. Here are a few examples.

Operation	*Example*	Mathematica *Command*
"+" Addition	$\begin{pmatrix} 3 & -4 & 7 \\ -1 & 0 & 5 \end{pmatrix} + \begin{pmatrix} 1 & 2 & 3 \\ 4 & 5 & 6 \end{pmatrix}$	**{{3,-4,7},{-1,0,5}} +** **{{1,2,3},{4,5,6}} //** **MatrixForm** $\begin{pmatrix} 4 & -2 & 10 \\ 3 & 5 & 11 \end{pmatrix}$
"−" Subtraction	$\begin{pmatrix} 3 & -4 \\ -1 & 0 \\ 7 & 5 \end{pmatrix} - \begin{pmatrix} 1 & 2 \\ 3 & 4 \\ 5 & 6 \end{pmatrix}$	**{{3,-4},{-1,0},{7,5}} -** **{{1,2},{3,4},{5,6}} //** **MatrixForm** $\begin{pmatrix} 2 & -6 \\ -4 & -4 \\ 2 & -1 \end{pmatrix}$
"*" Scalar multiplication	$5 \begin{pmatrix} 3 & -4 & 7 \\ -1 & 0 & 5 \end{pmatrix}$	**5*{{3,-4,7},{-1,0,5}} //** **MatrixForm** $\begin{pmatrix} 15 & -20 & 35 \\ -5 & 0 & 25 \end{pmatrix}$

| "." or "Dot" Matrix multiplication | $\begin{pmatrix} 3 & -4 & 7 \\ -1 & 0 & 5 \end{pmatrix} \cdot \begin{pmatrix} 2 & 0 & 1 & 5 \\ -1 & 1 & 2 & 0 \\ 3 & 5 & -2 & 1 \end{pmatrix}$ | `a = {{3,-4,7},{-1,0,5}}`
`b = {{2,0,1,5},{-1,1,2,0},`
`    {3,5,-2,1}}`
`MatrixForm[ a.b ]`
$\begin{pmatrix} 31 & 31 & -19 & 22 \\ 13 & 25 & -11 & 0 \end{pmatrix}$ |
| **MatrixPower** Power of a matrix | $\begin{pmatrix} 1 & 2 \\ 3 & 4 \end{pmatrix}^5$ | `a = {{1,2},{3,4}}`
`MatrixPower[a, 5] //`
`        MatrixForm`
$\begin{pmatrix} 1069 & 1558 \\ 2337 & 3406 \end{pmatrix}$ |

> **Note:** If **a** is a matrix, the expression **a^2** will only square each entry of the matrix. To get the square of the matrix, use **MatrixPower[a, 2]**.

Some Useful Matrix Commands

Mathematica has several commands that let you easily construct an identity matrix, as well as find the inverse, the determinant, the eigenvalues, and the eigenvectors of a given square matrix.

Here's a quick summary:

Operation	Mathematica *Command and Example*

IdentityMatrix – Make an identity matrix of a given size.	`IdentityMatrix[2]`
	`{{1, 0}, {0, 1}}`
	`IdentityMatrix[4] // MatrixForm`
	$\begin{pmatrix} 1 & 0 & 0 & 0 \\ 0 & 1 & 0 & 0 \\ 0 & 0 & 1 & 0 \\ 0 & 0 & 0 & 1 \end{pmatrix}$

Inverse – Find the inverse of a matrix.	`a = {{2, 5, 1}, {3, 1, 2}, {-2, 1, 0}};`
	`Inverse[a] // MatrixForm`
	$\begin{pmatrix} \dfrac{2}{19} & -\dfrac{1}{19} & -\dfrac{9}{19} \\ \dfrac{4}{19} & -\dfrac{2}{19} & \dfrac{1}{19} \\ -\dfrac{5}{19} & \dfrac{12}{19} & \dfrac{13}{19} \end{pmatrix}$

Det – Find the determinant of a matrix.	`a = {{2, 5, 1}, {3, 1, 2}, {-2, 1, 0}};`
	`Det[a]`
	`-19`

Eigenvalues – Find the eigenvalues of a matrix.	`b = {{1, 2, -1}, {2, 3, 1}, {1, 0, 2}};`
	`Eigenvalues[b]`
	$\left\{1, \dfrac{1}{2}\left(5 - \sqrt{13}\right), \dfrac{1}{2}\left(5 + \sqrt{13}\right)\right\}$

Eigenvectors
– Find the eigenvectors of a matrix.
	`b = {{1, 2, -1}, {2, 3, 1}, {1, 0, 2}};`
	`Eigenvectors[b]`
	
	$\{\{-2, 1, 2\}, \{\frac{1}{2} \left(1 - \sqrt{13}\right), \frac{1}{2} \left(5 - \sqrt{13}\right), 1\},$
	
	$\{\frac{1}{2} \left(1 + \sqrt{13}\right), \frac{1}{2} \left(5 + \sqrt{13}\right), 1\}\}$

Elementary Row Transformations

Row Vectors

Since a matrix is defined by listing its rows, you can easily retrieve and manipulate the rows of the matrix as vectors. You do this using double-bracket [[ ]] syntax.

```
a = {{2, 5, 1}, {3, 1, 2},{-2, 1, 0}};
MatrixForm[a]
```

$$\begin{pmatrix} 2 & 5 & 1 \\ 3 & 1 & 2 \\ -2 & 1 & 0 \end{pmatrix}$$

`a[[1]]`   (*This gives the 1st row.*)

`{2, 5, 1}`

`a[[3]]`   (*This gives the 3rd row.*)

`{-2, 1, 0}`

Row Operations

Recall from linear algebra that you often apply row transformations to a matrix to change its form into another, simpler form.

Mathematica can calculate row transformations for you, as you see in the following example.

■ **Example.** Consider the matrix $\begin{pmatrix} 2 & 1 & -1 & 1 \\ 1 & 0 & 3 & 4 \\ -5 & -3 & 1 & 2 \end{pmatrix}$.

```
a = {{2,1,-1,1}, {1,0,3,4}, {-5,-3,1,2}} ;
a // MatrixForm
```

$$\begin{pmatrix} 2 & 1 & -1 & 1 \\ 1 & 0 & 3 & 4 \\ -5 & -3 & 1 & 2 \end{pmatrix}$$

Step 1). We would like to interchange Row 1 and Row 2.

```
temp = a[[1]];
a[[1]] = a[[2]];
a[[2]] = temp;
```
          (*Makes a temporary copy of **a[[1]]** named **temp**.*)
          (*Redefines **a[[1]]** to be a copy of **a[[2]]**.*)
          (*Redefines **a[[2]]** to be **temp**, the original **a[[1]]**.*)

`a // MatrixForm`   (*You can see that Row 1 and Row 2 have been interchanged in this new matrix **a**.*)

$$\begin{pmatrix} 1 & 0 & 3 & 4 \\ 2 & 1 & -1 & 1 \\ -5 & -3 & 1 & 2 \end{pmatrix}$$

Step 2). We would like to replace Row 2 by "–2*Row 1 + Row 2"

```
a[[2]] = -2 a[[1]] + a[[2]];
a // MatrixForm
```

$$\begin{pmatrix} 1 & 0 & 3 & 4 \\ 0 & 1 & -7 & -7 \\ -5 & -3 & 1 & 2 \end{pmatrix}$$

Step 3). We would like to replace Row 3 by "5*Row 1 + Row 3"

```
a[[3]] = 5 a[[1]] + a[[3]];
a // MatrixForm
```

$$\begin{pmatrix} 1 & 0 & 3 & 4 \\ 0 & 1 & -7 & -7 \\ 0 & -3 & 16 & 22 \end{pmatrix}$$

Step 4). We would like to replace Row 3 by "3*Row 2 + Row 3"

```
a[[3]] = 3 a[[2]] + a[[3]];
a // MatrixForm
```

$$\begin{pmatrix} 1 & 0 & 3 & 4 \\ 0 & 1 & -7 & -7 \\ 0 & 0 & -5 & 1 \end{pmatrix}$$

Step 5). We would like to replace Row 3 by "–1/5 * Row 3"

```
a[[3]] = -1/5*a[[3]]
a // MatrixForm
```

$$\begin{pmatrix} 1 & 0 & 3 & 4 \\ 0 & 1 & -7 & -7 \\ 0 & 0 & 1 & -\frac{1}{5} \end{pmatrix}$$

Bingo!! We have transformed the original matrix into a nicer form. This is called the **row echelon form** in linear algebra.

Troubleshooting Q & A

Question...   I got an error message that "Objects of unequal length ... cannot be combined" when I tried to add vectors or matrices. What went wrong?

Answer...   This is the message you get when you try to add or subtract two vectors or matrices that have different shapes. You can only add or subtract vectors of the same length. Two matrices can be added or subtracted only if they have the same number of rows and the same number of columns.

Question... I tried to multiply two matrices but got the error message "Tensors ... and ... have incompatible shapes." What does that mean?

Answer... You can only compute the matrix product AB if the number of rows in B is the same as the number of columns in A. Check that the matrices you entered have compatible sizes.

Question... I got an error message when I used one of the **Det**, **Eigenvalues**, **Eigenvectors**, or **Inverse** commands. What should I check for?

Answer... Here are some suggestions:

- First, check that you entered the matrix correctly.

- Second, these commands only work for square matrices (same number of rows and columns), so check that too. Usually, the error message you'd get would tell you that your matrix "is not a square matrix."

- Also, the **Inverse** command can produce the error message "Linear equation encountered which has no solution." This means that the matrix is not invertible (its determinant is zero).

Question... I entered a row transformation incorrectly, but when I retyped the correct command, *Mathematica* still did not give the right answer. Why is that?

Answer... Each execution of a row transformation is done "in place," changing the matrix right away. For example, if you want to replace Row 3 by "2*Row 1 + Row 3," but instead you type:

```
a[[3]] = 5*a[[1]] + a[[3]];
```

The damage is already done! Row 3 has now been altered, and the matrix **a** has been changed. Even if you now type the correct expression:

```
a[[3]] = 2*a[[1]] + a[[3]];
```

this only makes the situation worse! You need to reenter the matrix and all the previous row transformations from the beginning. (Fortunately, this is very easy to do in the Notebook interface – just go back, reexecute the cell that defines **a**, and do the correct row transformations.)

CHAPTER 19

Parametric Curves and Surfaces in Space

ParametricPlot3D

Parametric Curves in Space

You use the **ParametricPlot3D** command to draw a space curve. To see the curve given parametrically as $(x(t), y(t), z(t))$, for $a \le t \le b$, type:

```
ParametricPlot3D[ {x(t),  y(t),  z(t)}, {t, a, b}]
```

This format is similar to the **ParametricPlot** command used for plane curves. Here, however, the curve is defined with three parametric functions rather than two.

■ **Example.** The helix given parametrically by $(t, 3\cos(t), 3\sin(t))$, for $0 \le t \le 8\pi$, is drawn with:

```
ParametricPlot3D[{t, 3Cos[t], 3Sin[t]}, {t,0,8Pi}]
```

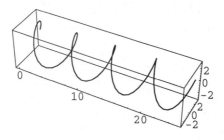

Options for Parametric-Plot3D

Most options you can use with **Plot3D** also work with **ParametricPlot3D**. In particular, **BoxRatios –> Automatic** is the default for the **ParametricPlot3D** command. That's why the graphic above is very wide and not very tall.

Also, you can remove the outside box and the axis labels by setting **Boxed –> False** and **Axes –> None**, respectively. Here's a nicer picture of the same helix.

```
ParametricPlot3D[{ t, 3Cos[t],3Sin[t]}, {t,0,8Pi},
       ViewPoint -> {2,1,1},
       BoxRatios -> {1,1,1},
       PlotPoints -> 150,
       Boxed -> False, Axes -> None]
```

Parametric Surfaces in Space

ParametricPlot3D can also be used to draw a surface in space. If a surface is defined parametrically by $(x(u,v), y(u,v), z(u,v))$ for $u_0 \le u \le u_1$ and $v_0 \le v \le v_1$, you enter:

```
ParametricPlot3D[{ x(u,v),  y(u,v),  z(u,v)},
                    {u,  u₀,  u₁}, {v,  v₀,  v₁ } ]
```

■ **Example.** To see a portion of the one-sheeted hyperboloid given parametrically by $(\cos(u)\cosh(v), \sin(u)\cosh(v), \sinh(v))$, for $0 \le u \le 2\pi$ and $-2 \le v \le 2$, write:

```
ParametricPlot3D[
   {Cos[u]Cosh[v],Sin[u]Cosh[v],Sinh[v]},
   {u, 0, 2Pi}, {v, -2, 2}]
```

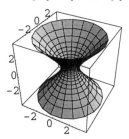

The surface $\left(v(2-\cos(4u))\cos(u), v(2-\cos(4u))\sin(u), v^2\right)$, for $0 \le u \le 2\pi$ and $0 \le v \le 2$, gives a very nice picture of a vase:

```
ParametricPlot3D[
   {v*(2-Cos[4*u])*Cos[u],v*(2-Cos[4*u])*Sin[u],v^2},
   {u,0, 2Pi}, {v, 0, 2},
     PlotPoints -> {60, 30}, ViewPoint -> {2,1,1}]
```

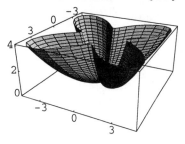

Plotting Multiple Curves and Surfaces

ParametricPlot3D can sketch several curves (or surfaces) with a single command, just like you've seen already with the **Plot** and **ParametricPlot** commands. Use one of these formats:

> **ParametricPlot3D[{ *curve1*, *curve2* }, {t, *a*, *b*}]**

or

> **ParametricPlot3D[{ *surface1*, *surface2* },**
> **{u, u_0, u_1}, {v, v_0, v_1 }]**

Multiple Curves

■ **Example.** The helixes $(t, 3\cos(t), 3\sin(t))$, and $(t, -3\sin(t), 3\cos(t))$ can be drawn together with:

```
ParametricPlot3D[
    { {t,3Cos[t],3Sin[t]}, {t,-3Sin[t],3Cos[t]} },
    {t, 0, 8Pi}, Boxed->False]
```

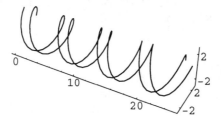

Multiple Surfaces

■ **Example.** The paraboloid $(r\cos t, r\sin t, r^2)$ opens up, while the paraboloid $(r\cos t, r\sin t, 2-r^2)$ opens down. We can combine them to form a nice "beehive."

```
ParametricPlot3D[ {{r*Cos[t],r*Sin[t],r^2},
    {r*Cos[t],r*Sin[t],2-r^2}},
    {r,0,1}, {t,0,2Pi}, ViewPoint -> {2,2,1}]
```

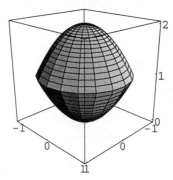

Styling Curves and Surfaces

Styles for Curves

Unlike the **Plot** and **ParametricPlot** commands, there is no **PlotStyle** option for **ParametricPlot3D**. Rather, you give a list of "style directives" as the *fourth coordinate of the parameterization* when you define the curve in the command.

Typical directives you might use are **Thickness**, **GrayLevel**, and **RGBColor**.

■ **Example.** To see the curve given parametrically by (t^3, t^2, t), for $0 \le t \le 2$ drawn in a thick, gray pen, you use:

```
style = {Thickness[0.03],GrayLevel[0.5]};
ParametricPlot3D[{t^3, t^2, t, style}, {t,0,2}]
```

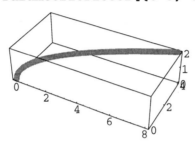

Shading for Surfaces

ParametricPlot3D illuminates a surface according to a default lighting scheme. This causes different portions of the surface to be colored or shaded differently.

You can turn off *Mathematica*'s default shading by setting **Shading –> False**. This will give an "opaque, plain" image. For example:

```
ParametricPlot3D[{2Cos[u],2Sin[u],v},
        {u, 0, 2Pi}, {v, 0, 5}, Shading->False ]
```

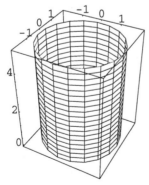

You can also control the shading more directly. However, the process is a little complicated, so we'll outline the procedure in three steps.

• First, you have to define a shading function using **FaceForm**.

```
shade[u_,v_] :=
    FaceForm[ instructions to shade the surface in terms of u, v ]
```

• Second, add the shading function as the *fourth* coordinate of the parameterization.

- Third, turn off *Mathematica*'s default illumination by setting **Lighting ->
  False**.

For example, suppose we want to shade the cylinder above according to its height, with darker shading at the bottom. We know the height, *v*, varies from 0 to 5. Our shading function should adjust the **GrayLevel** to vary from 0 to 1 accordingly:

```
shade[u_,v_] := FaceForm[GrayLevel[v/5]]
```

Now you can see the surface:

```
ParametricPlot3D[{2Cos[u], 2Sin[u], v, shade[u,v]},
    {u, 0, 2Pi}, {v, 0, 5}, Lighting->False]
```

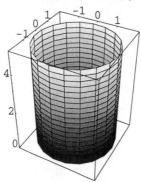

More Examples

Combining Graphics with Show

You can combine curves, surfaces, and any other three-dimensional images into a single graphic using the **Show** command (just as we saw in Chapter 15).

■ **Example.** The upper hemisphere of the unit sphere $x^2 + y^2 + z^2 = 1$ is given by ($r\cos(t), r\sin(t), \sqrt{1-r^2}$), for $0 \le r \le 1$ and $0 \le t \le 2\pi$.

```
g1 = ParametricPlot3D[
        {r*Cos[t], r*Sin[t], Sqrt[1-r^2]},
        {r, 0, 1}, {t, 0, 2Pi}]
```

The point $P = (\frac{1}{2}, \frac{1}{2}, \frac{1}{\sqrt{2}})$ lies on this hemisphere. A normal to the hemisphere at P is given by $\vec{r}(t) = (\frac{1}{2}+t, \frac{1}{2}+t, \frac{1}{\sqrt{2}}+\sqrt{2}\,t)$. This command shows just a portion of the normal line (notice that the fourth coordinate is a **Thickness** directive):

```
g2 = ParametricPlot3D[
        {1/2+t,1/2+t, 1/Sqrt[2]+Sqrt[2]*t,
        Thickness[0.02]}, {t, 0, 0.15}]
```

Finally, the plane tangent to the hemisphere at P has equation $z = \frac{2-x-y}{\sqrt{2}}$. You can sketch a portion of it near the point P with

```
g3 = Plot3D[(2-x-y)/Sqrt[2],
        {x,0.2,0.8},{y,0.2,0.8},PlotPoints->2]
```

You can now see one of the nicest features of *Mathematica* – the ability to combine these 3-D graphics despite the fact that each was drawn using a different type of command.

Show[g1, g2, g3, ViewPoint -> {3,-1,.5}]

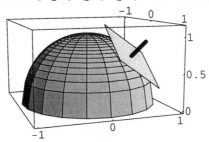

Notice how well the graphic convinces you that we have described both the tangent plane and the normal line correctly!

Troubleshooting Q & A

Question... When I tried to draw a curve in space, I got an error message "Function ... cannot be compiled," and then got an empty 2-D frame box. What happened?

Answer... The most common mistake is to forget that **ParametricPlot3D** is used for a space curve, while **ParametricPlot** is used for a plane curve. If you use **ParametricPlot**, but enter the formula for a space curve, you will get an empty 2-D frame box.

Question... I entered a command to draw a parametric surface, but *Mathematica* returned the command to me without drawing any picture. What went wrong?

Answer... There are many possibilities, but our best suggestions are:

- Did you use **ParametricPlot3D**? A common mistake is trying to draw a parametric surface with either the **ParametricPlot** or **Plot3D** command.

- Check that you have entered the command correctly. Be especially careful with the syntax of the parameterization.

Question... When I used **ParametricPlot3D**, I got an error message "... is not a list starting with three real numbers." What happened?

Answer... This usually means that *Mathematica* cannot properly evaluate the three coordinate functions that you input for a surface or a curve. Check:

- Did you enter the function(s) correctly without a typing mistake? Is each of the functions defined everywhere in the interval(s) you specified?

- Did you use the same literal parameter in both your function and interval? (For example, check that you don't write **f[x,y]** when the parameters are **u** and **v**.)

- Did you use two parameters for a surface?

Question... I tried to use a **ParametricPlot3D** command for a surface and got the message "`Limiting value ... is not a machine-size real number.`" What should I check for?

Answer... Check that you used numbers when you specified the endpoints of the intervals in the command. In particular, the endpoints for one variable cannot depend on the other variable.

For example, the upper hemisphere of the unit sphere has equation $z = \sqrt{1 - x^2 - y^2}$, for $-1 \le x \le 1$, and $-\sqrt{1-x^2} \le y \le \sqrt{1-x^2}$. But the following doesn't work:

```
ParametricPlot3D[{x,y,Sqrt[1-x^2-y^2]},
    {x,-1,1}, {y,-Sqrt[1-x^2],Sqrt[1-x^2]}]
```

This does not work because the interval specified for y depends on the other variable x. You should look for a different parameterization to see the hemisphere. (Hint: Try cylindrical or spherical coordinates!)

Question... I used **ParametricPlot3D** and got the message "`Function ... cannot be compiled.`" Then, I got the picture I was expecting anyway. Should I be worried about this?

Answer... No. Consider these two methods to show part of a paraboloid:

```
ParametricPlot3D[ {u,v,u^2+v^2}, {u,-2,2}, {v,-2,2} ]
```

and:

```
s[u_,v_] := {u, v, u^2+v^2 }
ParametricPlot3D[ s[u,v], {u,-2,2}, {v,-2,2} ]
```

The first uses the parameterization **{u,v,u^2+v^2}** directly in the command. When *Mathematica* sees this, it can use a computer trick (called *compilation*) to speed up drawing the surface.

The second method uses **s[u,v]** to define the surface. Since this doesn't look like a list, *Mathematica* cannot use compilation to make the command run as fast. This is what generates the warning message, but you can ignore it. And you won't notice the speed difference, either.

CHAPTER 20
Vector Fields

Drawing a Vector Field

PlotVector-Field Command

The **PlotVectorField** command sketches vector fields in the plane. It is defined in the package **Graphics`PlotField`**. To see the vector field defined by $\vec{F}(x, y) = (F_1(x, y), F_2(x, y))$ for $x_0 \le x \le x_1$, $y_0 \le y \le y_1$, you enter:

```
Needs["Graphics`PlotField`"]
PlotVectorField[ { F_1(x, y) , F_2(x, y) },
                  { x, x_0, x_1 }, { y, y_0, y_1 }]
```

For example, to see the vector field $(-y, x)$, for $-2 \le x \le 2$, $-2 \le y \le 2$, you type:

```
Needs["Graphics`PlotField`"]
PlotVectorField[{-y,x}, {x,-2,2},{y,-2,2}]
```

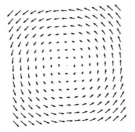

> **Note:** It may take awhile for *Mathematica* to show the picture above!

Strictly speaking, *Mathematica* does not draw the vector field $(-y, x)$ exactly. It has scaled the lengths of the vectors proportionally to produce a nice picture.

Some Useful Options

Some options for the **PlotVectorField** command can help us change the style of the output picture.

Option	What It Does
PlotPoints -> 8	Draws $8 \times 8 = 64$ vectors.
Axes -> Automatic	Shows the axes.
ScaleFunction -> (1&)	Draws all vectors with the same length.

For example, here's a different picture of the vector field above:

```
PlotVectorField[{-y,x}, {x,-2,2},{y,-2,2},
        PlotPoints -> 5, Axes -> Automatic,
        ScaleFunction -> (1&)]
```

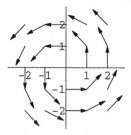

Vector Fields in Space

To draw a 3-D vector field (in space), you use the **PlotVectorField3D** command. It is defined in the **Graphics`PlotField3D`** package.

To see the vector field $\vec{F}(x, y, z) = (F_1(x, y, z), F_2(x, y, z), F_3(x, y, z))$, for the intervals $x_0 \le x \le x_1$, $y_0 \le y \le y_1$, and $z_0 \le z \le z_1$, you type:

```
Needs["Graphics`PlotField3D`"]
PlotVectorField3D[{ F_1(x, y, z),  F_2(x, y, z),  F_3(x, y, z)},
                {x,  x_0,  x_1}, {y,  y_0,  y_1}, {z,  z_0,  z_1}]
```

For example,

```
Needs["Graphics`PlotField3D`"]
PlotVectorField3D[{-z,1,x}, {x,-2,2},
                        {y,-2,2}, {z,-2,2}]
```

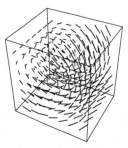

This looks like a mess. Let's try a **ViewPoint** looking from the positive, y-axis:

```
Show[%, ViewPoint -> {0.2,2,0.7}]
```

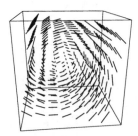

It looks like the vector field is swirling around the y-axis.

Gradient, Curl, and Divergence

Gradient Field   The gradient of a function is denoted by grad f. It is defined to be the vector

$$\text{grad } f = (f_x, f_y), \text{ or grad } f = (f_x, f_y, f_z),$$

depending on whether f is a function of two or three variables, respectively. You can find the gradient of f by calculating the partial derivatives explicitly:

```
Clear[f]
f[x_,y_,z_] := x^2 + 2x*y - z^3
{D[f[x,y,z], x], D[f[x,y,z],y], D[f[x,y,z],z]}
```
$$\left\{2\,x + 2\,y,\ 2\,x,\ -3\,z^2\right\}$$

You can also use the **Grad** command from the **Calculus`VectorAnalysis`** package. This lets you find the gradient in any standard coordinate system (e.g., cartesian, spherical, or cylindrical, and some esoteric ones too!).

The following table gives examples of computing the gradient in the three most-used coordinate systems. But first, we load the package that defines the **Grad** command:

```
Needs["Calculus`VectorAnalysis`"]
```

Coordinate System	Mathematica *Command*

Cartesian or rectangular (x, y, z)	`Grad[x^2 + 2x*y - z^3,`
	`    Cartesian[x,y,z]]`
	
	$\left\{2\,x + 2\,y,\ 2\,x,\ -3\,z^2\right\}$

Cylindrical (r, θ, z)	`Grad[r*Cos[theta]+r^2*z,`
	`    Cylindrical[r,theta,z]]`
	
	$\left\{2\,r\,z + \text{Cos[theta]},\ -\text{Sin[theta]},\ r^2\right\}$

Spherical (ρ, θ, ϕ)	`Grad[2rho^2*Cos[theta]*Sin[phi],`
	`    Spherical[rho, theta, phi]]`
	
	{4 rho Cos[theta] Sin[phi],
	-2 rho Sin[phi] Sin[theta],
	2 rho Cos[phi] Cot[theta]}

> **Note:** The order of the variables in **Cartesian[x,y,z]**, **Cylindrical[r,theta,z]**, and **Spherical[rho, theta, phi]** is important. If you list them in a different order, the computation of the gradient will be incorrect.

Curl and Divergence   To calculate the curl and divergence of a vector field, use the **Curl** and **Div** commands, respectively. They are also defined in the **Calculus`VectorAnalysis`** package and are used much as the **Grad** command. First load them with:

```
Needs["Calculus`VectorAnalysis`"]
```

Examples	Mathematica *Command*
Divergence in Cartesian system	`Div[{x^2,y^2,z^2}, Cartesian[x,y,z]]` 2 x + 2 y + 2 z

Curl in Cartesian system	`Curl[ {x+y, y+z, Sin[x*y]+z^2},` `Cartesian[x,y,z]]`
	$\{-1 + x\,\text{Cos}[x\,y], -y\,\text{Cos}[x\,y], -1\}$
Divergence in spherical system	`Div[{rho^2, rho*Sin[phi], Sin[theta]},` `Spherical[rho, theta, phi]]`
	$\dfrac{1}{\text{rho}^2}\left(\text{Csc[theta]}\left(\text{rho}^2\,\text{Cos[theta]}\,\text{Sin[phi]} +\right.\right.$ $\left.\left. 4\,\text{rho}^3\,\text{Sin[theta]}\right)\right)$

Line and Surface Integrals

Integration of a Vector Field

Engineers and physicists are often interested in integrating a vector field $\vec{F}$ either along a curve $\vec{r}(t)$, for $a \le t \le b$ or over a surface $\vec{s}(u,v)$, for $u_0 \le u \le u_1$, $v_0 \le v \le v_1$. These are defined as:

- Line integral: $\displaystyle\int_a^b \vec{F}(\vec{r}(t)) \cdot \vec{r}'(t)\, dt$

- Surface integral: $\displaystyle\pm\int_{u_0}^{u_1}\int_{v_0}^{v_1} \vec{F}(\vec{s}(u,v)) \cdot \left(\frac{\partial \vec{s}}{\partial u} \times \frac{\partial \vec{s}}{\partial v}\right) dv\, du$ (The choice of $\pm$ sign depends on how the normal for the surface is defined.)

These integrals are easily computed using the **Integrate** command, as shown in the following examples.

Line Integral

■ **Example.** Let $\vec{F}$ be the vector field $\vec{F}(x, y) = (x + y, -y)$ and $\vec{r}(t)$ the parametric curve $\vec{r}(t) = (1 - t, t^2)$ for $0 \le t \le 1$. The line integral of $\vec{F}$ over this curve is computed as:

```
F[{x_,y_}] := { x+y, -y }      (*This defines the vector field F.*)
r[t_] := { 1-t, t^2 }

Integrate[ F[r[t]] . r'[t], {t, 0, 1}]
```

$$-\frac{4}{3}$$

> **Note:** Syntax is important! We defined **r[t_]** to give a 2-D vector and **F[{x_,y_}]** to define the vector field $\vec{F}$ above. This lets us compute $\vec{F}(\vec{r}(t))$ easily with the expression **F[r[t]]**.

We can see why the line integral turned out to be negative by checking how the vector field lines up with the curve.

```
Needs["Graphics`PlotField`"]
field = PlotVectorField[ F[{x,y}], {x,0,1},{y, 0, 1}]
curve = ParametricPlot[ r[t], {t, 0, 1}]
Show[ curve, field, AspectRatio -> Automatic]
```

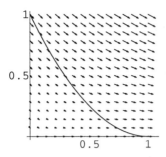

As you can see from the picture, if a particle moves along the curve from its starting point $\vec{r}(0) = (1,0)$ to its ending point $\vec{r}(1) = (0, 1)$, it is moving against the force of the vector field. Thus, the line integral is negative.

Surface Integral

■ **Example.** Suppose the surface integral of $\vec{F}(x, y, z) = (x(1 + z), y, 0)$ over the surface S is given by

$$-\int_1^4 \int_0^{2\pi} \vec{F}(\vec{s}(u, v)) \cdot \left(\frac{\partial \vec{s}}{\partial u} \times \frac{\partial \vec{s}}{\partial v} \right) dv \, du$$

where S is parameterized by $\vec{s}(u, v) = (u\cos v, u\sin v, u)$. We can evaluate this integral as follows:

```
F[{x_,y_,z_}] := {x(1+z), y, 0}
s[u_,v_] := {u*Cos[v], u*Sin[v], u}

su = D[s[u,v], u];        (*This calculates ∂s̄ / ∂u.*)
sv = D[s[u,v], v];        (*This calculates ∂s̄ / ∂v.*)

Needs["LinearAlgebra`CrossProduct`"]

- Integrate[F[s[u,v]] . Cross[su, sv],
        {v, 0, 2*Pi}, {u, 1, 4}]
```

$$\frac{423 \, \pi}{4}$$

More Examples

The Perpendicular Property of the Gradient

■ **Example.** Consider the function $f(x, y) = xy + 2x$:

```
f[x_,y_] := x*y +2 x
```

The perpendicular property of the gradient vector states that grad $f(a,b)$ is perpendicular to the level curve of f that goes through (a,b). To see this, we first draw several level curves:

```
pict1 = ContourPlot[f[x,y], {x,-4,4}, {y,-4,4},
            ContourShading->False, Contours->20]
```

By direct computation, grad $f(x, y) = (y + 2, x)$. We can draw this gradient field with the **PlotVectorField** command.

```
Needs["Graphics`PlotField`"]
pict2 = PlotVectorField[{y+2,x}, {x,-4,4}, {y,-4,4},
                    ScaleFunction -> (1&)]
```

Now combine the contours and gradient vectors into one picture:

Show[pict1, pict2]

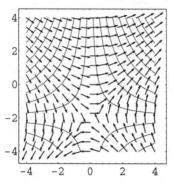

Do you agree that the gradient vectors are perpendicular to the level curves at every point?

Useful Tips

Instead of loading each of the individual sub-packages **Graphics`PlotField`** and **Graphics`PlotField3D`**, you can load all of the **Graphics** packages with:

Needs["Graphics`Master`"]

This way, you do not have to remember the exact name of the package that contains **PlotVectorField** or **PlotVectorField3D**.

Troubleshooting Q & A

Question... I did not get a picture from **PlotVectorField** or **PlotVectorField3D**. What should I check for?

Answer... There are several possible mistakes.

Input errors:

- Did you load the package, **Graphics`PlotField`** or **Graphics`PlotField3D`**? If you forgot, type the following and then reload the package:

Remove[PlotVectorField,PlotVectorField3D]

- Did you mistakenly use **PlotVectorField** for a 3-D vector field, or **PlotVectorField3D** for a 2-D vector field?

- Check that you have input the correct format and expression.

Hardware problems:

- Does the computer have enough memory to draw the picture (especially if you have already drawn a lot of graphics)? Restarting *Mathematica* may help.

- If your computer is slow, it can take a long time to show the output. Reducing the number of vectors by using the **PlotPoints** option can dramatically improve speed.

Question... When I used **Grad**, **Curl**, or **Div**, I got the wrong answer 0 or {0,0,0}. What was my mistake?

Answer... You forgot to specify the coordinate system and the variable names in the command. You have to type **Cartesian[x, y, z]**, **Cylindrical[r, theta, z]**, **Spherical[rho, theta, phi]**, and so on.

If you don't say otherwise, version 3.0 of *Mathematica* assumes that you're using rectangular coordinates with variables named **Xx**, **Yy**, and **Zz**. Version 2.2 had other default settings that often caused confusion.

We recommend that you always specify the coordinate system and variable names when using **Grad**, **Curl**, and **Div**.

Question... I defined a vector field named **F** and then defined a curve named **r[t_]** (or a surface **s[u_,v_]**), but when I put these together to calculate the line integral (or surface integral), *Mathematica* returned the integration unevaluated. What should I check?

Answer... First look at the output. Does it have the characters "F," "r," or "s" in it?

If it does, this means that *Mathematica* does not understand your definition of **F**, **r** or **s**. Check that you have defined the vector field in the format **F[{x_,y_}]** or **F[{x_,y_,z_}]**. Also check that the vector field has the same dimensions as your curve (or surface).

If the output does not contain the characters "F," "r," or "s", it suggests that *Mathematica* has trouble with the integration. Replace **Integrate** with **NIntegrate** to get a numerical answer.

CHAPTER 21
Basic Statistics

Well over one hundred specialized statistical and data analysis commands are available in *Mathematica*. In this chapter and the next, we will show you some basic statistics tools that you can use. Once you are familiar with the underlying concepts, the use of other advanced statistics tools is generally straightforward.

Graphical Presentation of Data

Scatter Diagrams, Pie or Bar Charts

A set of numerical data can sometimes be presented graphically by using a scatter diagram, a pie chart, or a bar chart. You do this with the commands **ListPlot**, **PieChart**, and **BarChart**, respectively.

The **ListPlot** command is best used in the format:

```
ListPlot[ data , PlotStyle -> PointSize[0.02]]
```

(See Chapter 6 for more information on **ListPlot**.) The **PieChart** and **BarChart** commands are defined in an external package and must be loaded before they can be used. Both commands have the same syntax:

```
Needs["Graphics`Graphics`"]
BarChart[ data ]
PieChart[ data ]
```

■ **Example.** Here are data from the UNESCO 1990 *Demographic Yearbook* which are pairs {*male life expectancy, female life expectancy*} in the countries of Argentina, Austria, Afghanistan, Algeria, Angola, Bolivia, Brazil, Belgium, Bangladesh, and Botswana, respectively.

```
data = { {65.5,72.7}, {73.3,79.6}, {41,42},
         {61.6,63.3}, {42.9,46.1}, {51,55.4},
         {62.3,67.6}, {70,76.8},   {56.9,56},
         {52.3,59.7}}
```

```
ListPlot[data, PlotStyle->PointSize[0.02],
         PlotLabel->"Life expectancy {Male, Female}"]
```

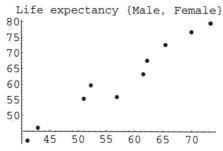

126

Notice that these data suggest a "linear relation" between male and female life expectancies. We will investigate this further in the next chapter.

■ **Example.** Here are the home run totals for North American baseball teams of the National League for the 1997 season, accompanied by abbreviated city names.

```
natLeagueHomeRuns =
   {{174, At}, {239, Co}, {142, Ci}, {127, Ch},
    {136, Fl}, {133, Ho}, {174, LA}, {172, Mo},
    {153, NY}, {129, Pi}, {116, Ph}, {152, SD},
    {172, SF}, {144, SL}};
```

A bar or pie chart shows at a glance which teams were most productive.

```
Needs["Graphics`Graphics`"]
BarChart[natLeagueHomeRuns]
```

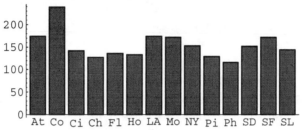

```
PieChart[natLeagueHomeRuns]
```

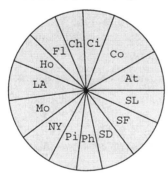

Numerical Measures

Mean, Median, Standard Deviation, etc.

Standard statistical measures of data distribution are available in *Mathematica*. To use them, you first have to load the master statistics package:

```
Needs["Statistics`Master`"]
```

Let us again consider the number of home runs for baseball teams in the National League for the 1997 season, which we used in the previous example.

```
homeRuns =  {174, 239, 142, 127, 136, 133, 174,
             172, 153, 129, 116, 152, 172, 144};
```

(You can also extract these data directly from **natLeagueHomeRuns** in the previous example, using

```
homeRuns = Map[First, natLeagueHomeRuns];
```

The use of the **Map** and the **First** commands will be explained in detail in Chapter 24, "More about Lists.")

We can find the mean and median directly with the following commands.

Mean[homeRuns] // N   (*We use **// N** to see the numerical value.*)

154.5

Median[homeRuns] // N

148

Other measures of central tendency such as the **HarmonicMean**, **GeometricMean** and **RootMeanSquare** are also available and work in same manner.

The best-known measures of data variability also have their customary names:

Variance[homeRuns]   // N

960.115

StandardDeviation[homeRuns]   // N

30.9857

MeanDeviation[homeRuns] // N

22.6429

These data can help us measure how closely the observations adhere to a specific statistical model. Say, if the data are drawn from a normal distribution, the command **MeanCI** will give the 95% confidence interval for the mean.

MeanCI[homeRuns] // N

{136.609, 172.391}   (*This tells us that we expect the actual mean to lie between 136.609 and 172.391, with a confidence level of 95%.*)

Similarly, **VarianceCI** computes a 95% confidence interval for the variance. By changing the value of its **ConfidenceLevel** option, we can find, say, a 90% confidence interval for the variance:

VarianceCI[homeRuns,ConfidenceLevel->0.90] // N

{558.156,2118.43}

Probability Distributions

The standard statistics packages contain definitions of common probability distributions. These include the following, along with their usual parameter specifications:

NormalDistribution[μ, σ]	**UniformDistribution**[a, b]	**BinomialDistribution**[λ]
StudentTDistribution[n]	**PoissonDistribution**[λ]	**BetaDistribution**[α, β]
ExponentialDistribution[λ]	**ChiDistribution**[n]	**ChiSquareDistribution**[n]

Cumulative Distribution and Probability Density Functions

You can work with the **cumulative distribution function** and the **probability density function** of each of these distributions using the commands:

CDF[*distribution* **, x]**   (*Cumulative distribution function.*)

PDF[*distribution*, **x]**   (*Probability density function.*)

For example, consider the normal distribution with mean 10 and deviation 3.

```
Needs["Statistics`Master`"]
distribution1 = NormalDistribution[10,3]
```

To find the values of its cumulative distribution function and probability density function, say at $x = 15$, use:

CDF[distribution1,15]   (*Cumulative distribution function at 15.*)

0.95221

PDF[distribution1,15]   (*Probability density function at 15.*)

0.033159

Also, we can see the graphs of the cumulative distribution and probability density functions, respectively, with:

Plot[CDF[distribution1, x], {x,0, 20}]

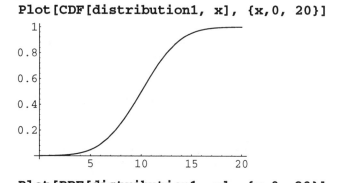

Plot[PDF[distribution1, x], {x,0, 20}]

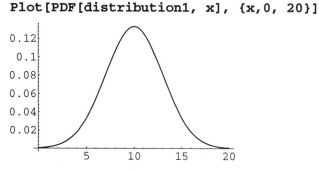

You can work with other distributions in a similar way. All you need is to include the proper parameter(s) when you reference the distributions.

More Examples

IQ score   ■ **Example**: It is generally believed that IQ scores are normally distributed with mean 100 and standard deviation 15. Let us theoretically estimate how many people in the world have IQs higher than 180.

The probability that a person's IQ is between 0 and 180 is:

```
CDF[NormalDistribution[100, 15], 180] // N
```

0.99999995

So the probability that a person's IQ is higher than 180 is:

```
1 - %
```

4.8213×10^{-8}

This probability can also be obtained by integrating the probability density function for $x \geq 180$.

```
NIntegrate[PDF[NormalDistribution[100,15], x],
                        {x, 180, Infinity}]
```

4.8213×10^{-8}

There are about 5 billion people in this world, so:

```
% * 5*10^9
```

241.065

Only about 250 people in this world have IQs higher than 180. Isn't it odd, then, how often we meet people who claim to have such high IQs?

Generating a Random Sample from a Specified Distribution

■ **Example**. Using the commands **Random** and **Table**, you can generate a random sample from specified distributions. For example, to obtain 50 random values from the Poisson distribution with mean 5, we use:

```
prandom :=  Random[PoissonDistribution[5]]
list1 = Table[prandom, {50}]
```

```
{6, 2, 10, 3, 6, 3, 5, 4, 5, 6, 6, 5, 3, 5, 3, 3, 9,
   5, 8, 5, 9, 8, 6, 5, 3, 3, 5, 3, 4, 2, 1, 4, 1, 4,
   4, 6, 6, 7, 5, 4, 7, 5, 3, 5, 5, 6, 7, 9, 3, 5}
```

The mean of this sample is:

```
Mean[list1] // N
```

4.94   (*This is pretty close to the theoretical value of 5.*)

Let us consider **ExponentialDistribution[3]**. Its mean and variance can be found with :

```
Mean[ExponentialDistribution[3]]
```

$\dfrac{1}{3}$

```
Variance[ExponentialDistribution[3]]
```

$\dfrac{1}{9}$

Now, let us generate a sample of 1000 random observations from this distribution. We can see that the mean and variance of this random sample are very close to the theoretical values of $\frac{1}{3}$ and $\frac{1}{9}$, respectively:

```
randomdata := Random[ExponentialDistribution[3]]
list2 = Table[randomdata, {1000}];
Mean[list2]
```

0.316677

```
Variance[list2]
```

0.112533

**Importing
Numerical
Values from a
File**

You can easily import data values into *Mathematica* from another application (e.g., a spreadsheet or an e-mail message) and then use the statistical tools we've introduced in this chapter to study the data.

For example, suppose you have a data file named "**TestScores.dat**" where individual data values are written in text format, one to a line. You can read all the data values into a list named **dataValues** by using the following commands:

```
channel = OpenRead["TestScores.dat"];
dataValues = ReadList[channel];
Close[channel];
```

Now you can work with and analyze the data in the list **dataValues** by using commands from *Mathematica*'s statistics packages.

We will explain more about the **OpenRead** and **ReadList** commands in Chapter 26.

Useful Tips

 When you need to use a statistics command, we recommend that you load the entire collection of statistics packages with:

```
Needs["Statistics`Master`"]
```

You can certainly load each individual statistics package separately. But, if you use this command, you're less likely to run into the usual troubles associated with loading the wrong package.

Troubleshooting Q & A

Question ... I forgot to load the statistics packages when I first used a statistics command. After I realized it, I loaded the statistics packages, but still the commands did not work. What went wrong?

Answer ... Before you reload the packages, you should **Remove** all the names of the commands that you tried to use before loading. For example:

```
Remove[Mean]
Needs["Statistics`Master`"]
```

Question ...   When I typed in the name of a distribution, *Mathematica* returned the expression to me without doing anything. Is that what's supposed to happen?

Answer ...   If you have already loaded the statistics packages, then it is normal that *Mathematica* returned the distribution command to you. For example,

`NormalDistribution[500, 20]`

`NormalDistribution[500, 20]`

The name of a distribution is used only to identify its particular distribution. By itself, it does not produce any output. You have to use it with **PDF**, **CDF**, **Random**, and so on, in order to see any sort of result. For example,

`CDF[NormalDistribution[500,20], 527]`

`0.911492`

Question ...   When I generated a bar chart, the *x*- and *y*-coordinates were interchanged. What went wrong?

Answer ...   This was not your fault. In generating bar charts, *Mathematica* treats pairs in the format {*y*-coordinate, *x*-coordinate}. Thus the first coordinate of each pair determines the bar's height! This is different from the standard convention.

A way you can quickly fix the problem is to use the **BarChart** command with each of the data pairs reversed. Use this syntax:

`mylist` = *the data pairs that you've defined as x-coordinate, y-coordinate*

`BarChart[Map[Reverse,mylist]]`

See Chapter 24 for more information about the **Map** and **Reverse** commands.

CHAPTER 22

Regression and Interpolation

Regression

The Fit Command

Given a list of data, you may want to find the line that "best" fits this data set. (Best fit is determined by the "least square method".) This process is called **linear regression**. We can use the following form of the **Fit** command to do this.

```
Fit[ data, {1, x}, x]
```

■ **Example.** Let us revisit data reported in the UNESCO 1990 *Demographic Yearbook* that we introduced at the start of the previous chapter. It consists of data pairs {*male life expectancy, female life expectancy*} in the countries of Argentina, Austria, Afghanistan, Algeria, Angola, Bolivia, Brazil, Belgium, Bangladesh, and Botswana, respectively.

```
data = {  {65.5,72.7},  {73.3,79.6},  {41,42},
          {61.6,63.3},  {42.9,46.1},  {51,55.4},
          {62.3,67.6},  {70,76.8},    {56.9,56},
          {52.3,59.7}}
```

Previously, we had only displayed the data using **ListPlot**. We'll do so again, but this time we'll first compute the line that best fits the data.

```
bestfit[x_] = Fit[data, {1, x}, x]
```

```
-3.19208 + 1.12885 x
```

We can see how well the line fits the data with:

```
pict1 = ListPlot[data, PlotStyle -> PointSize[0.02]]
pict2 = Plot[bestfit[x], {x,40,75}]
Show[pict1, pict2]
```

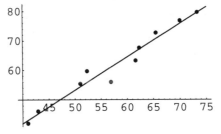

Under this model, say, if a certain country's male life expectancy is 65 years, then the female life expectancy is estimated to be:

```
bestfit[65]
```

```
70.1832
```

The Regress Command

You can also use the command **Regress** (defined in the statistics packages) to find the best fitting line. In addition, this command will give a full analysis of the accuracy of the fit. Using our previous example:

```
Needs["Statistics`Master`"]
Regress[data,{1, x}, x]
```

{ParameterTable →

	Estimate	SE	TStat	PValue
1	-3.19208	4.81367	-0.663129	0.52589
x	1.12885	0.0821508	13.7412	7.58796×10^{-7}

,

RSquared → 0.959354,

AdjustedRSquared → 0.954273, EstimatedVariance → 7.18389, ANOVATable →

	DF	SumOfSq	MeanSq	FRatio	PValue
Model	1	1356.46	1356.46	188.82	7.58796×10^{-7}
Error	8	57.4712	7.18389		
Total	9	1413.94			

}

You can see that the RSquared value is very close to 1. This means that a strong correlation exists between the life expectancy of males and females in the countries.

Fit, Regress, and Other Functions

The **Fit** and **Regress** commands can be used with functions other than just 1 and x to provide different types of least-square fits to data. Given functions $f_1(x)$, $f_2(x)$, $f_3(x)$, ..., $f_n(x)$, if you want to find a function of the form

$$a_1 f_1(x) + a_2 f_2(x) + a_3 f_3(x) + ... + a_n f_n(x)$$

that best fits a given set of data, you will use:

Fit[*data,* **{** $f_1(x), f_2(x), f_3(x), ..., f_n(x)$ **},** **x]**

For example, using the previous data:

Fit[data, {1, x, x^2}, x]   (*Best fit quadratic function.*)

$5.74472 + 0.802964 \ x + 0.00287186 \ x^2$

Fit[data, {1, x, x^2,x^3}, x]   (*Best fit cubic function.*)

$-122.129 + 7.8101 \ x - 0.122245 \ x^2 + 0.000729509 \ x^3$

Fit[data, {1, Exp[2x], Exp[x],Exp[-x],Sin[x]},x]

$0. + 6.61047 \times 10^{-144} \ E^{-x} + 2.60284 \times 10^{-94} \ E^x +$
$1.71361 \times 10^{-62} \ E^{2x} - 7.99794 \times 10^{-127} \ Sin[x]$

Interpolation

The Interpolating-Polynomial Command

Given n points (x_1, y_1), (x_2, y_2), ..., (x_n, y_n), you can find a polynomial of degree $n - 1$ that perfectly matches all these values. This can be done in *Mathematica* using the command **InterpolatingPolynomial** in the syntax:

```
InterpolatingPolynomial[ data, x ]
```

For example, to find a polynomial $g(x)$ that passes through all of the points, use:

```
data = { {0,0},    {1,16},   {2,-10}, {3,28},
         {4,-30}, {5,-12}, {6,-12}};
```

```
g[x_] = InterpolatingPolynomial[data, x];
Expand[g[x]]
```

$$\frac{13733\,x}{30} - \frac{162041\,x^2}{180} + \frac{1897\,x^3}{3} - \frac{1819\,x^4}{9} + \frac{299\,x^5}{10} - \frac{299\,x^6}{180}$$

You can check that $g(x)$ passes through all the given points with:

```
Table[{x,g[x]}, {x,0,6}]
```

```
{{0,0}, {1,16}, {2,-10}, {3,28}, {4,-30}, {5,-12},
  {6,-12}}
```

You can see that the graph of $g(x)$ passes through all the given points with:

```
pict1 = ListPlot[data, PlotStyle->PointSize[0.03]]
pict2 = Plot[g[x], {x, 0,6}]
Show[pict1, pict2]
```

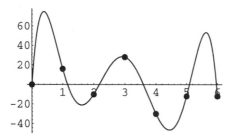

> **Note:** Make sure you use "= "and not ":=" when defining a function in terms of an **InterpolatingPolynomial**.

More Examples

Fit Vs Interpolating-Polynomial

■ **Example.** You may wonder about the difference between regression (**Fit**) and interpolation (**InterpolatingPolynomial**). Regression allows us to find a function of a specified form that best approximates the given set of data. On the other hand, interpolation will find a function that matches the data exactly. But such a function may behave wildly, as you see in the picture of the previous example.

Consider the UNESCO data of the first example of this chapter.

```
data = {  {65.5,72.7}, {73.3,79.6}, {41,42},
          {61.6,63.3}, {42.9,46.1}, {51,55.4},
          {62.3,67.6}, {70,76.8},   {56.9,56},
          {52.3,59.7}}
```

If we had used interpolation rather than finding a least-square fit for this data, we would have found the polynomial:

```
perfectfit[x_] = InterpolatingPolynomial[data, x];
Expand[perfectfit[x]] // N
```

$$2.03962 \times 10^8 - 3.36365 \times 10^7\, x + 2.45307 \times 10^6\, x^2 -$$
$$103838.\, x^3 + 2811.63\, x^4 - 50.5044\, x^5 + 0.601845\, x^6 -$$
$$0.00458829\, x^7 + 0.0000203073\, x^8 - 3.97572 \times 10^{-8}\, x^9$$

Let's compare graphically the results obtained using regression (shown in gray) and interpolation:

```
bestfit[x_] = Fit[data, {1, x}, x];
pict1 = Plot[bestfit[x], {x,40,75}, PlotStyle ->
                {{Thickness[0.01], GrayLevel[0.7]}}]
pict2 = Plot[perfectfit[x], {x, 40,75}]
pict3 = ListPlot[data, PlotStyle -> PointSize[0.02]]
```

```
Show[pict1, pict2, pict3]
```

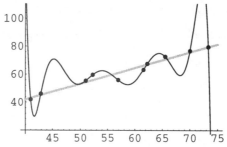

You can see in this case that regression (**Fit**) is likely to be better for prediction.

A Temperature Model

■ **Example.** The following monthly data in the form { *month, temperature in* $F°$ } show the average monthly high temperatures in Boston, Massachusetts (1 represents the month of January, 2 for February, and so on).

```
bostonHigh = {  {1, 36.4}, {2, 37.7}, {3,  45.0},
                {4,  56.6}, {5, 67.0}, {6,  76.6},
                {7,  81.8}, {8, 79.8}, {9,  72.3},
                {10, 62.5},{11,51.6},  {12,40.3},
                {13, 36.4},{14, 37.7},{15, 45.0}};
g1 = ListPlot[bostonHigh, PlotStyle->PointSize[0.02]]
```

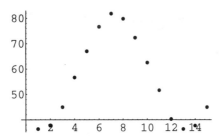

Since temperature is cyclic, we expect that it can be described nicely by a cosine function in the form

$$Temperature = a_1 + a_2 \cos(a_3(x - a_4)), \text{ where } x \text{ is the month.}$$

The temperature's period is 12 months, so we should choose $a_3 = \frac{2\pi}{12} = \frac{\pi}{6}$. Also, since we expect July to be the hottest month of the year, the function must have its largest value when $x = 7$. This means that we should choose $a_4 = 7$.

Thus, we need to find the function of the form $a_1 + a_2 \cos(\frac{\pi}{6}(x - 7))$ that best fits the temperature data:

```
fTemp[x_] = Fit[bostonHigh, {1,Cos[Pi/6(x-7)]}, x];
g2 = Plot[fTemp[x],{x,-3,15}];
Show[g1,g2]
```

The temperature data fit nicely in this model. (Actually, the result can be even more impressive if we choose $a_4 = 7.25$ instead. Try it yourself!)

A U.S. Population Model

■ **Example.** The U.S. population (measured in millions) over the last two hundred years is given as follows:

```
pop =
  { {1790,   3.929}, {1800,   5.308}, {1810,   7.240},
    {1820,   9.638}, {1830,  12.861}, {1840,  17.063},
    {1850,  23.192}, {1860,  31.443}, {1870,  38.558},
    {1880,  50.189}, {1890,  62.980}, {1900,  76.212},
    {1910,  92.228}, {1920,106.021}, {1930,123.203},
    {1940,132.166}, {1950,151.326}, {1960,179.323},
    {1970,203.302}, {1980,226.542}, {1990,248.710}};
```

Theoretically, this population growth will follow a "logistic model" and hence has the form:

$$\text{population} = \frac{288.5}{1 + e^{a+bx}}, \text{ where } x \text{ is the year}$$

(The constant 288.5 represents the maximum sustainable population 288.5 million, which can be predicted by the data.) We want to find the constants a and b so that the logistic model best fits the data.

We cannot use the **Fit** command directly to find a and b. However, one can change the expression to the form:

$$\text{population} = \frac{288.5}{1 + e^{a+bx}} \quad \Leftrightarrow \quad \frac{288.5}{\text{population}} = 1 + e^{a+bx}$$

$$\Leftrightarrow \quad \ln(\frac{288.5}{\text{population}} - 1) = a + bx$$

This suggests that we can find a and b using the new data $\ln(\frac{288.5}{\text{population}} - 1)$. Thus we must transform each of the given data pairs $\{x, y\}$ to be $\{x, \ln(\frac{288.5}{y} - 1)\}$.

```
f[{x_,y_}] := {x, Log[288.5/y -1]}
newdata = Map[f, pop]
```

```
{{1790, 4.2826}, {1800, 3.97691}, {1810, 3.65966},
 {1820, 3.365}, {1830, 3.06489}, {1840, 2.76682},
 {1850, 2.43708}, {1860, 2.10112}, {1870, 1.86907},
 {1880, 1.55778}, {1890, 1.27559}, {1900, 1.02442},
 {1910, 0.755238}, {1920, 0.542998}, {1930, 0.29391},
 {1940, 0.167936}, {1950, -0.0981863},
 {1960, -0.496218},  {1970, -0.869715},
 {1980, -1.29647}, {1990, -1.83267}}
```

(The **Map** command will be explained in Chapter 24.)

```
Fit[newdata, {1, x}, x]
```

```
55.3245 - 0.0285529 x
```

This suggests that $a = 55.3245$ and $b = -0.0285529$. The logistic model will thus be:

```
g[x_] := (288.5)/ (1 + Exp[55.3245-0.0285529*x])
pict1 = ListPlot[pop, PlotStyle -> PointSize[0.02]]
pict2 = Plot[g[x], {x,1790,1990}]
Show[pict1, pict2]
```

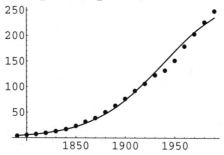

The logistic model fits the data very nicely, especially through about 1930.

Useful Tips

☼   If you are planning to do a lot of statistics, we recommend that you also read Chapters 24 and 25, to learn more about lists, random numbers and simulation.

Troubleshooting Q & A

Question ...   I used the **Fit** command, but got an error message stating that the "Number of coordinates ... is not equal to the number of parameters". What went wrong?

Answer ...   Your input data has to be of the form,

$$\{\{x_1, y_1\}, \{x_2, y_2\}, \ldots, \{x_n, y_n\}\}$$

or

$$\{y_1, y_2, y_3, \ldots, y_n\}$$

(In the second case, *Mathematica* will automatically assume that $x_1 = 1$, $x_2 = 2$, $x_3 = 3$, and so on.) If your input has more than two coordinates, say,

```
data = { {1,2,3}, {2,4,6}, {3,1,6}, {2,1,1} }
```

then you have to specify that there are two parameters x and y.

```
Fit[data, {1, x, y}, {x, y}]
```

Question ...   When I tried to use a function that I defined using the **InterpolatingPolynomial** command, I got the error message "... cannot be used as the variable in a polynomial." What went wrong?

Answer ...   When you define a function using **InterpolatingPolynomial**, you should use:

```
h[x_] = InterpolatingPolynomial, {data, x}]
```

Do not use ":=" to define a function in terms of an **InterpolatingPolynomial**.

Question ...   Can you explain the difference between using the **Fit** and **InterpolatingPolynomial** commands?

Answer ...   **Fit** finds a function of a specified form that best *approximates* the given set of data. On the other hand, **InterpolatingPolynomial** will find a function that *matches the data exactly*, although it may behave "wildly." See the first example of the "More Examples" section.

CHAPTER 23

Animation

Getting Started

What Is Animation?

Let your eyes scan the 15 graphic frames shown below. You should be able to see a point P "move" along the sine curve, just as if you were watching a movie.

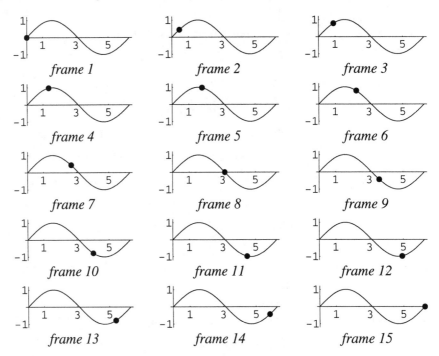

You can ask *Mathematica* to create many static pictures (like the ones you see above) called *frames* and then display them rapidly enough so that your eye sees a continuous motion.

The Animate Command

It is fun and easy to do animation in *Mathematica*. You use the **Animate** command in the following form (the **Needs** command can be omitted if you are using version 3.0 or higher of *Mathematica*):

```
Needs["Graphics`Animation`"]
Animate[ a graphics command involving t , {t, t_0 , t_1 }]
```

The *graphics command* inside **Animate** can be **Plot**, **ParametricPlot**, **Plot3D**, etc., that we discussed in earlier chapters. It will be executed to produce frames of the animation for numerous values of t between t_0 and t_1. For example, with

```
Needs["Graphics`Animation`"]
Animate[ Plot[Sin[t*x],{x,0,2Pi}], {t,1,10}]
```

Mathematica produces a sequence of frames that are approximately the same as the output of the commands:

```
Plot[Sin[1*x],{x,0,2Pi}]          (*frame 1*)
Plot[Sin[1.4*x],{x,0,2Pi}]        (*frame 2*)
(* etc. *)
```

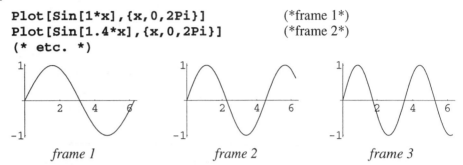

frame 1   *frame 2*   *frame 3*

After *Mathematica* finishes creating all the pictures, you can double-click on any one of the frames; then the animation will begin!

Animation control buttons will appear at the bottom, left corner of your Notebook window when the animation begins. You can click on these as indicated to adjust how the animation plays, just as you would control a VCR.

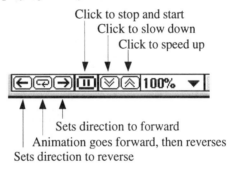

Animating a 3-D Picture

You can make 3-D animations in *Mathematica* as well. You only have to use a 3-D graphic command inside **Animate**. For example,

```
Needs["Graphics`Animation`"]
Animate[ Plot3D[Sin[t*x]+Cos[t*y],{x,0,Pi},{y,0,Pi}],
    {t,1,5}]
```

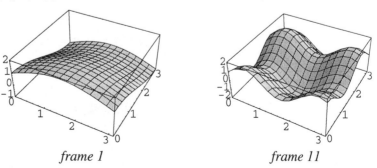

frame 1   *frame 11*

We show only two frames of the animation above (frames 1 and 11). If you watch the animation, you'll see "waves" move through the surface.

More Examples

**Writing
Animations**

The key to writing sophisticated animations is to have a firm grasp of what each frame looks like, as well as how the frames vary according to a parameter. Following this concept, you can easily create a complicated animation.

In each of the following examples, we'll define **oneFrame**, a "frame function" of t, that will generate the graphic elements for each frame. Because each frame itself will be a combination of several simple graphics, we'll also include all the graphics commands inside (*parentheses*):

> **oneFrame[t_] := (** *graphics command(s) depending on t* **)**

Then we can use the **Animate** command, sometimes adding the **Frames** option, which lets us specify exactly how many frames will be produced for the animation:

```
Needs["Graphics`Animation`"]
Animate[ oneFrame[t] , {t, t0 , t1 }, Frames -> n ]
```

> **Note**: The default number of frames produced in an animation is 24. You may want to increase this to see smoother animations. You may also want to lower this for complicated animations if your computer does not have much memory.

**Moving a Point
Along a Curve**

■ **Example.** Let's show you how to write the animation of a point moving along the sine graph, $y = \sin x$ (the one you saw on the first page of this chapter). The co-ordinate of the point at any time t can be thought as $P(t) = (t, \sin(t))$ for $0 \leq t \leq 2\pi$.

In each frame of the animation, we will plot the position of the point at time t against a background of the sine curve. The sine curve can be sketched with a **Plot** command, and the point can be drawn using a **ListPlot** command. We combine them using **Show**.

The graphic outputs of **Plot** and **ListPlot** must be turned off by setting **DisplayFunction –> Identity** when they are drawn individually. Later, graphic output must be restored by setting **DisplayFunction –> $DisplayFunction** when we use the **Show** command. Here's our frame function:

```
Clear[oneFrame]
oneFrame[t_]:=
    (g1 = Plot[Sin[x], {x,0,2Pi},
          DisplayFunction->Identity];
     g2 = ListPlot[{{t,Sin[t]}},
          PlotStyle->PointSize[0.03],
          DisplayFunction->Identity];
     Show[g1, g2, AspectRatio->Automatic,
          DisplayFunction->$DisplayFunction] )
```

For example, you can see the frame at $t = 0.5$ with:

```
oneFrame[0.5]
```

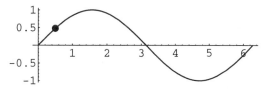

Now we ask *Mathematica* to **Animate** over the interval $0 \leq t \leq 2\pi$ using 15 *frames*.

```
Needs["Graphics`Animation`"]
Animate[ oneFrame[t], {t,0,2Pi}, Frames->15 ]
```

Rolling a Ball Along the Ground

■ **Example**. A circle of radius one "rests" on top of the *x*-axis at the origin, as pictured below. A point *P*, which is marked at the "top" of the circle, is currently touching the *y*-axis at the point (0, 2).

As the circle begins to "roll" to the right, the point *P* will rotate downward and eventually hit the *x*-axis after the circle has rolled a distance of π (because *P* was originally halfway around a circle of circumference 2π). Thereafter, *P* will rotate upward again after the circle "rolls over" it, reaching a high point again when the circle has rolled a distance of 2π.

The following picture suggests the motion of *P*. The path of *P* is marked in gray.

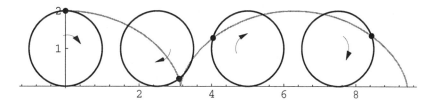

In fact, the path followed by *P* as the circle rolls is known to be a cycloid given parametrically by $(u + \sin(u), 1 + \cos(u))$, for $-\infty < u < \infty$.

We will let the parameter *t* in this animation be the distance the ball has rolled. Each frame of the animation should have three elements:

- A fixed portion of the cycloid to act as a background. It can be given by $(u + \sin(u), 1 + \cos(u))$, for $0 \leq u \leq 4\pi$.

- The new position of the unit circle in this frame. Since the circle has moved a distance *t*, its center is at $(t, 1)$ and its equation is thus $(t + \cos(u), 1 + \sin(u))$.

- The corresponding position of the point *P*. After the ball has traveled a distance of *t*, *P* will have rotated a distance of *t* clockwise around the circle. *P*'s location relative to the center of the circle is thus given by $(\sin(t), \cos(t))$. Its position in the frame will then be $(t + \sin(t), 1 + \cos(t))$.

After some experimentation with **PlotRange**, our frame function is as follows.

```
Clear[oneFrame]
oneFrame[t_] :=
    (g1 = ParametricPlot[{u+Sin[u],1+Cos[u]},
            {u,0,4Pi}, PlotStyle->GrayLevel[0.75],
            DisplayFunction->Identity];
    g2 = ParametricPlot[{t+Cos[u],1+Sin[u]},
            {u,0,2Pi}, DisplayFunction->Identity];
    g3 = ListPlot[{{t+Sin[t],1+Cos[t]}},
            PlotStyle->PointSize[0.02],
            DisplayFunction->Identity];
    Show[g1,g2,g3,
            DisplayFunction->$DisplayFunction,
            AspectRatio->Automatic,
            PlotRange->{{-3,15},{-.1,2.1}}]  )
```

The completed animation which rolls P over two arcs of the cycloid is given with:

```
Needs["Graphics`Animation`"]
Animate[ oneFrame[t], {t,0,4Pi} ]
```

The Oscillating Spring

■ **Example.** A metal plate is attached to a spring of height h_0. (The picture to the right shows such a spring at a height of 5 units above the xy-plane.) The plate is then displaced upward a distance of c units and released. The plate (and spring) will begin a vertical, damped oscillation.

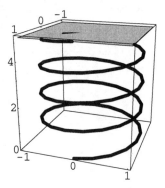

If we ignore the effect of gravity and the weight of the plate, then the height of the plate at any time t after release will be

$$z = h_0 + c\,e^{-bt}\cos(kt),$$

where b and k are constants that depend on the spring.

We'll model the motion of the plate using animation in *Mathematica*. The graphic elements of each frame are as follows.

- The height of the "plate" is $z = h_0 + c\,e^{-bt}\cos(kt)$. It can be seen with **Plot3D**. We'll sketch it only over the rectangle $-1 \le x \le 1$, $-1 \le y \le 1$.

- We'll think of the "spring" as the helix $(\cos(u),\ \sin(u),\ \frac{u}{8\pi}(h_0 + c\,e^{-bt}\cos(kt)),$ for $0 \le u \le 8\pi$, and sketch it using **ParametricPlot3D**.

Here's the frame function:

```
Clear[oneFrame, h0, c, b, k]
oneFrame[t_] :=
    (g1 = Plot3D[h0+c*Exp[-b*t]Cos[k*t],
            {x,-1,1}, {y,-1,1}, PlotPoints->2,
            PlotRange->
                {Automatic,Automatic,{0,h0+c+0.1}},
            DisplayFunction->Identity];
    g2 = ParametricPlot3D[{Cos[u],Sin[u],
                u*(h0+c*Exp[-b*t]Cos[k*t])/(8Pi),
                Thickness[0.02]},
            {u,0,8Pi},
            DisplayFunction->Identity];
    Show[g1,g2,
            DisplayFunction->$DisplayFunction,
            ViewPoint->{3,1,1},
            BoxRatios->{1,1,2}])
```

Before you can see an animation, you must choose numerical values for the various constants h_0, c, b, and k we've introduced. For example:

```
h0 = 5;       (*The plate is at height 5 at rest.*)
c = 2;        (*It is displaced 2 units upward.*)
b = 0.05;     (*This coefficient controls damping.*)
k = 1;        (*This coefficient controls speed of oscillation.*)
```

Now you can do the animation, say, over the time interval $0 \le t \le 8$:

```
Animate[oneFrame[t],{t,0,8}]
```

Try repeating the animation for different numerical values of the constants and see how each one affects the movement of the plate. For example, increasing the value of b will cause the oscillation to die out more quickly.

Useful Tips

🔆 🔆 🔆 🔆 If your frame function uses **Show** to combine several graphics, make sure you set **DisplayFunction –> Identity** when you draw each of the graphics. Later, be sure to set **DisplayFunction –> $DisplayFunction** when you use the **Show** command.

🔆 🔆 🔆 To get the best results of an animation, you must line up the individual frames of the animation exactly. Make sure you explicitly set the **PlotRange** in the frames so that you have a consistent set of frames.

Troubleshooting Q & A

Question... **Animate** doesn't seem to work at all. What should I check?

Answer... Prior to version 3.0 of *Mathematica*, you had to explicitly load the **Animate** command from an external package before using it. If you didn't do this before trying your animation, you should clear up the definition of **Animate** with:

```
Remove[Animate]
Needs["Graphics`Animation`"]
```

Next, make sure you test out your frame function before you use it with the **Animate** command. For example:

```
oneFrame[t_] := ( your frame function definition )
```

```
oneFrame[0]
```

Did you get the right picture when $t = 0$?

Question...   The **Animate** command drew some pictures, but I couldn't see a real animation happen. What should I do?

Answer...   Once the pictures are drawn, you must double-click one of the frames to animate it. (There's also a menu command you can use to get the animation started.)

On a rare occasion, your animation might be running OK, but you don't see any real motion because the frames of your animation are all the same! For example:

```
Animate[ Plot[t*Sin[x], {x,-Pi,Pi}] , {t,1,4} ]
```

Using this command, you'll get the same frame every time because, even though the graphs of $f(x) = t\sin(x)$ are different, they will all be drawn using the same aspect ratio. The only things that will be different from frame to frame are the labels on the y-axis.

Question...   I got **Animate** to work, but I didn't get exactly the type of animation I expected. The animation was not very smooth, and a few of the frames even seemed to "jump" around. What should I do?

Answer...   Make sure you've set up an explicit **PlotRange** in your frame function. This makes sure that all your frames have consistent measurements, with axes placed in the same position in every frame.

Question...   *Mathematica* crashed after it drew a few frames of an animation. What should I do?

Answer...   Animations use lots of memory. Version 3.0 of *Mathematica* handles memory a little better than version 2.2, but both will likely die on complicated animations (especially 3-D).

Try increasing the memory allocation for *Mathematica*. Closing unneeded windows and quitting other applications may help too.

Also, try your animation using a small value for **Frames**. Gradually increase the number of frames until you get a satisfactory result.

CHAPTER 24

More About Lists

In this chapter we will show you more commands that you can use when you work with lists. Many of them will be helpful in statistics or *Mathematica* programming.

Displaying Lists

ColumnForm and TableForm

To display lists nicely, use one of the **ColumnForm** and **TableForm** commands. For example, consider the list:

```
myList = Table[ 2^n , {n, 1, 5}]
```

```
{2, 4, 8, 16, 32}
```

The **ColumnForm** command arranges the elements line by line:

```
ColumnForm[myList]
```

```
2
4
8
16
32
```

You can also use its postfix syntax:

```
myList // ColumnForm
```

```
2
4
8
16
32
```

If each element of a list is itself a list, then you should use the **TableForm** command to line up the output in both rows and columns. For example:

```
myList2  = Table[ {n,n^2,2^n} , {n, 1, 4}]
```

```
{{1, 1, 2}, {2, 4, 4}, {3, 9, 8}, {4, 16, 16}}
```

```
TableForm[ myList2 ]
```

```
1    1     2
2    4     4
3    9     8
4    16    16
```

Displaying Long Lists

When your list is too long to display on screen, you'll find it helpful to suppress its output with a semicolon. (You do not want the whole screen filled up with numbers.)

> `lotsofsquares = Table[n^2,{n,1,2000}];`

Then you can use either the **Short** or **Shallow** command to see part of the list.

> `Short[lotsofsquares]`
>
> `{1, 4, 9, 16, 25, <<1992>>, 3992004, 3996001, 4000000}`

(*The term `<<1992>>` means 1992 items have been suppressed in the output.*)

> `Shallow[lotsofsquares]`
>
> `{1, 4, 9, 16, 25, 36, 49, 64, 81, 100, <<1990>>}`

As a second example, it might be of interest to construct a list of the first 1000 prime integers, using the **Prime** operator. We find these with:

> `lotsofprimes = Table[ Prime[i], {i,1,1000} ];`
> `lotsofprimes // Short`
>
> `{2, 3, 5, 7, 11, 13, 17, 19, 23, <<987>>, 7883, 7901,`
> `    7907, 7919}`

You were probably already familiar with the first few primes, but not many people realize that 7919 is the 1000-th prime.

Useful List Commands

The Count Command	The **Count** operator shows how many times a particular item appears in a list. It has the syntax:

> **Count** [*your list* , *particular item*]

For example,

> `list1 = { 3, 3, 5, 6, 5, 9, 10, 5, 5, 1, 5, 1, 5}`
>
> `{3, 3, 5, 6, 5, 9, 10, 5, 5, 1, 5, 1, 5}`
>
> `Count[ list1, 3 ]`
>
> `2`   (*3 appears 2 times in **list1**.*)
>
> `Count[ list1, 5 ]`
>
> `6`   (*5 appears 6 times in **list1**.*)

The Map Command	To evaluate a function at each element of a list, you use the **Map** command. It has the form:

> **Map** [*function name* , *the list*]

Let's look at an easy example:

> `list1 = Table[ 2^n, {n, 1, 8}]`
>
> `{2, 4, 8, 16, 32, 64, 128, 256}`
>
> `f1[x_] := x + 1`   (*We want to add 1 to each element.*)
> `Map[ f1, list1]`

```
{3, 5, 9, 17, 33, 65, 129, 257}
```

> **Note:** Inside the **Map** command, we type the *name* of the function only. In the example above, we use **f1** and not **f1[x]** .

Now, we'll try a more interesting example:

```
list2 = {  {"John", 75, 62}, {"David", 62, 81},
           {"Mary", 75, 91}, {"Jane",  31, 50},
           {"Steve",21, 31}}
```

This shows a list of five students and their scores in two exams. We will calculate the average score of each student and list it with their name:

```
f2[{name_, score1_, score2_}] :=
          {name, N[(score1+score2)/2]}
```

```
Map[f2, list2]
```

```
{{John, 68.5}, {David, 71.5}, {Mary, 83.},
 {Jane, 40.5}, {Steve, 26.}}
```

The **Map** command provides an easy way to implement "data transformation" or "data massaging" if you work with statistical data. Consult Chapter 22's example on modeling the US population for this.

Other List Commands

Many other commands for lists will be useful, especially for programming in *Mathematica*. We will list some of them here. All of the following examples will be demonstrated using one or both of the next two lists.

```
list3 = { 2, 5, -3, 1.2 , 6.01, 7.5}
list4 = { 1.2, 8, 7.5, 3}
```

Mathematica *Command*	*Explanation*
`list3[[5]]` `6.01`	**[[** *n* **]]** will pick out the *n-th* element of the list.
`Length[list3]` `6`	**Length** tells you the number of elements in a list.

`Max[list3]`
`7.5`
`Min[list3]`
`-3`
	Min and **Max** are used to find the largest and smallest elements of a list, respectively.

`First[list3]`
`2`
`Last[list3]`
`7.5`
	First and **Last** are used to find the first and last elements of a list.

`Take[list3, {2, 5}]` `{5, -3, 1.2, 6.01}`	Here, we use the **Take** command to pick out the 2-nd up to the 5-th elements.
`Drop[list3, 3]` `{1.2, 6.01, 7.5}`	Here, we use the **Drop** command to remove or "drop" the first three elements of the list.

`Drop[list3, -2]` `{2, 5, -3, 1.2}`	When used with a negative number, **Drop** removes or "drops" elements off the end of the list.
`Sort[list3]` `{-3, 1.2, 2, 5, 6.01, 7.5}`	**Sort** will arrange the elements of the list in an increasing order.
`Reverse[list3]` `{7.5, 6.01, 1.2, -3, 5, 2}`	**Reverse** will reverse the arrangement of the list. That is, the first element becomes the last and vice versa.
`Union[list3, list4]` `{-3, 1.2, 2, 3, 5, 6.01, 7.5, 8}`	**Union** combines the elements of both lists, removes duplicate elements, and sorts the result.
`Intersection[list3, list4]` `{1.2, 7.5}`	**Intersection** finds the elements that are common to both lists, removes duplicate elements and sorts the result.

Calculating the Average

■ **Example**. It's easy to compute the average of the elements of a list using **Sum**, **Length**, and the [[*n*]] notation. For example, to find the average of the elements in **list3**:

```
list3 = { 2, 5, -3, 1.2 , 6.01, 7.5}
```

```
Sum[ list3[[i]], {i,1,Length[list3]} ] / Length[list3]
```
```
3.11833
```

That is, the average is just the sum of the elements, divided by the number of elements in the list. Of course, you can also use the **Mean** command from the Statistics packages to compute this. (See Chapter 21.)

More Examples

Baseball Standings

■ **Example**. The following list represents the final standings of the American League Eastern Division Baseball teams for the 1997 regular season in the form {*team*, *wins*, *losses*}:

```
baseball = { {"Bal", 98, 64}, {"Bos", 78, 84},
   {"Det", 79, 83}, {"NY", 96, 66}, {"Tor", 76, 86}};
```

```
TableForm[baseball]
```
```
Bal    98    64
Bos    78    84
Det    79    83
NY     96    66
Tor    76    86
```

When you look at the sports page of a newspaper, you'll see that the team standings include the team's winning percentage. Also, teams with a higher winning percentage will be listed first. Let's show how to do this in *Mathematica*:

- We will first add the winning percentage to each element (team):

```
f1[{team_, wins_, losses_}] :=
        { N[wins/162, 3], team, wins, losses }

list1 = Map[ f1, baseball ]
```

```
{{0.605, Bal, 98, 64}, {0.481, Bos, 78, 84},
   {0.488, Det, 79, 83}, {0.593, NY, 96, 66},
   {0.469, Tor, 76, 86}}
```

- Next we will use **Sort** to arrange the elements in increasing order by winning percentage, and then we will use **Reverse** to make it a decreasing order instead:

```
list2 = Reverse[Sort[list1]]
```

```
{{0.605, Bal, 98, 64}, {0.593, NY, 96, 66},
   {0.488, Det, 79, 83}, {0.481, Bos, 78, 84},
   {0.469, Tor, 76, 86}}
```

- Finally, we will rearrange each element to have the form { *team*, *wins*, *losses*, *winning percentage* } and show it in **TableForm**:

```
f2[{pct_, team_, wins_, losses_}] :=
            {team, wins, losses, pct }
Map[f2, list2] // TableForm
```

```
Bal    98    64    0.605
NY     96    66    0.593
Det    79    83    0.488
Bos    78    84    0.481
Tor    76    86    0.469
```

The Digits of Pi

■ **Example**. We all know that π is an irrational number. Thus, the digits in its decimal expansion do not repeat and behave unpredictably. We can see the first 100 digits of π with:

```
N[ Pi, 100 ]
```

```
3.141592653589793238462643383279502884197169399375102\
5820974944592307816406286208998628034825342117068
```

You may wonder how many times each of the digits 0, 1, 2, ..., 9 shows up in the expansion above. To find out, we first need to convert the digits into a list. This is done with the **RealDigits** command:

```
RealDigits[N[Pi,100]]
```

```
{{3, 1, 4, 1, 5, 9, 2, 6, 5, 3, 5, 8, 9, 7, 9, 3,
   2, 3, 8, 4, 6, 2, 6, 4, 3, 3, 8, 3, 2, 7, 9,
   5, 0, 2, 8, 8, 4, 1, 9, 7, 1, 6, 9, 3, 9, 9,
   3, 7, 5, 1, 0, 5, 8, 2, 0, 9, 7, 4, 9, 4, 4,
   5, 9, 2, 3, 0, 7, 8, 1, 6, 4, 0, 6, 2, 8, 6,
   2, 0, 8, 9, 9, 8, 6, 2, 8, 0, 3, 4, 8, 2, 5,
   3, 4, 2, 1, 1, 7, 0, 6, 8}, 1}
```

The result is a list of 2 elements. The first element is a list itself consisting of 100 digits of π. The second (last) element is the number 1, indicating that *one* digit is ahead of the decimal point.

```
list1 = %[[1]]; Short[list1]
```

{3, 1, 4, 1, 5, 9, 2, 6, <<88>>, 7, 0, 6, 8}

Now, we will count how many times each of the digits $0, 1, \ldots, 9$ appears in this list:

```
data = Table[ {Count[list1, n], n}, {n, 0, 9}]
```

{{8, 0}, {8, 1}, {12, 2}, {12, 3}, {10, 4},
    {8, 5}, {9, 6}, {7, 7}, {13, 8}, {13, 9}}

This means that 0 appears 8 times, 1 appears 8 times, 2 appears 12 times, and so on. A better way to look at this is to use the **BarChart** command.

```
Needs["Graphics`Graphics`"]
BarChart[data]
```

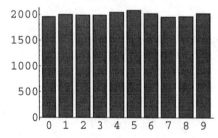

It looks like the numbers 8 and 9 appear more often in the first 100 digits of π. Repeating the steps above, but adjusting them for 20,000 digits of π, you will see a different result.

```
list2 = RealDigits[ N[Pi, 20000] ] [[1]] ;
BarChart[ Table[{Count[list2, i], i}, {i,0,9}]]
```

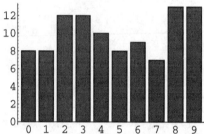

A Game of Bridge, Anyone?

■ **Example**. What happens if you want to play bridge, but you don't have a deck of cards? We can ask *Mathematica* to simulate shuffling a deck of cards.

A deck consists of 52 cards, arranged into four suits named "Clubs," "Diamonds," "Hearts," and "Spades" of 13 cards each. Asking *Mathematica* to shuffle cards is mathematically the same as arranging the numbers from 1 to 52 randomly. We can use the **RandomPermutation** command to do this:

```
Needs["DiscreteMath`Permutations`"]
list1 = RandomPermutation[52]
```

```
{3, 49, 29, 28, 16, 34, 18, 11, 50, 43, 12, 22,
 13, 40, 1, 9, 26, 14, 27, 36, 51, 23, 46, 52,
 10, 7, 44, 8, 48, 37, 32, 30, 42, 5, 6, 25, 20,
 47, 35, 41, 15, 45, 17, 38, 33, 19, 4, 31, 24,
 2, 39, 21}
```

Now we need to interpret the cards numbered 1 to 13 as "Spades," those numbered 14 to 26 as "Hearts," 27 to 39 as "Diamonds," and 40 to 52 as "Clubs."

```
f1[x_] := Which[   1 <= x <= 13, {"S", x},
                  14 <= x <= 26, {"H", (27-x)},
                  27 <= x <= 39, {"D", (40-x)},
                  40 <= x <= 52, {"C", (53-x)}]

list2 = Map[f1, list1];
Shallow[list2]
```

```
{{S, 3}, {C, 4}, {D, 11}, {D, 12}, {H, 11}, {D, 6}, <<46>> }
```

It will help to interpret 1 as "Ace," 11 as "Jack," 12 as "Queen," and 13 as "King":

```
f2[{x_,y_}] := Which[ y == 1,          {x, "Ace"},
                      2 <= y <= 10,    {x, y},
                      y == 11,         {x, "Jack"},
                      y == 12,         {x, "Queen"},
                      y == 13,         {x, "King"}   ]

list3 = Map[f2, list2];
```

Finally, we give the first 13 cards to South, the next 13 cards to West, the next 13 cards to North and the remaining cards to East.

```
south =  Sort[Take[ list3, {1, 13}]]
```

```
{{C, 3}, {C, 4}, {C, 10}, {D, 6}, {D, Jack}, {D, Queen},
 {H, 5}, {H, 9}, {H, Jack}, {S, 3}, {S, Jack}, {S, Queen}}
```

```
west =  Sort[Take[list3, {14, 26}]]
```

```
{{C, 2}, {C, 7}, {C, Ace}, {C, King}, {D, 4}, {D, King},
 {H, 4}, {H, Ace}, {H, King}, {S, 7}, {S, 9}, {S, 10},
 {S, Ace}, {S, King}}
```

```
north =  Sort[Take[list3, {27,39}]]
```

```
{{C, 5}, {C, 6}, {C, 9}, {C, Jack}, {D, 3}, {D, 5},
 {D, 8}, {D, 10}, {H, 2}, {H, 7}, {S, 5}, {S, 6}, {S, 8}}
```

```
east = Sort[Take[list3, {40,52}]]
```

```
{{C, 8}, {C, Queen}, {D, 2}, {D, 7}, {D, 9}, {D, Ace},
 {H, 3}, {H, 6}, {H, 8}, {H, 10}, {H, Queen}, {S, 2}, {S, 4}}
```

It looks like West has a pretty good hand!

CHAPTER 25

Random Numbers and Simulation

Random Numbers

The Random Command

We can use *Mathematica* to generate sequences of random numbers. Such sequences can be used to simulate probabilistic situations very easily.

To obtain a random real number between 0 and 1, you type:

```
Random[]
```

Each time you execute **Random[ ]**, you get a different result. For example:

```
Random[]
0.0211633
```

```
Random[]
0.722949
```

You can generate either integer- or real-valued random numbers between given numbers *a* and *b* with:

```
Random[ Integer, { a, b }]
```

or

```
Random[ Real, { a, b }]
```

For example:

```
Random[ Real, {1.2, 8.5}]   (* Gives a real between 1.2 and 8.5. *)
5.47079
```

```
Random[ Integer, {3, 8}]    (* Gives an integer between 3 and 8. *)
7
```

Pseudo-Randomness

Sequences of numbers returned by **Random** are said to be "pseudo-random." They are actually generated using an iteration formula. They eventually repeat, but not soon enough so that you'd notice (in fact, you'd have to go about 10^{445} iterations before you'd see any repetition).

And, as you see below, the values of **Random** certainly "look" random!

```
data = Table[Random[],{1000}];        (*Generates 1000 random
                                         numbers between 0 and 1.*)

ListPlot[data]                        (*This shows the numbers.*)
```

The fact that there seems to be no obvious pattern emerging confirms our sense of randomness.

The Distribution of Random

■ **Example.** Let us generate 5000 random integers between 1 and 10:

```
data1 = Table[ Random[Integer,{1, 10}], {5000}];
```

You can see some of them with:

```
Short[data1]

{1, 4, 7, 4, 10, 7, 7, 6, 6, <<4986>>, 4, 2, 9, 9, 8}
```

If these values are really "random," we expect each of the numbers 1, 2, . . . , 10 will show up about the same number of times. Let's check it:

```
data2 = Table[{Count[data1,n], n}, {n, 1, 10}]

{{506, 1}, {511, 2}, {487, 3}, {501, 4}, {518, 5},
  {473, 6}, {518, 7}, {504, 8}, {493, 9}, {489, 10}}
```

So, 1 appeared 506 times, 2 appeared 511 times, and so forth. In fact, each digit appeared almost equally often (close to 500 times). Consequently, we say that **Random** produces a uniform distribution of numbers between 0 and 1.

You can see this even better visually:

```
Needs["Graphics`Graphics`"]
BarChart[ data2 ]
```

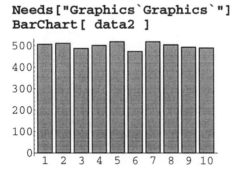

Examples in Simulation

With the help of the **Random** command and a little programming, we can simulate many real-world experiments in *Mathematica*. In each of the following simulations, usually three procedures are involved.

- We will write down the procedure(s) that generates the result of a single experiment. In many cases the procedure involves the **If** command, which has the syntax:

  If[*condition, result1, result2***]**

  Mathematica will give *result1* if *condition* is True. If *condition* is False, *result2* is returned. (The **If** command is discussed in the next chapter.)

- Then, we use the **Table** command to repeat the experiment a large number of times in the form:

  Table[*one result of a random experiment,* **{** *n* **}]**

- After we obtain experimental data in this way, we'll analyze the data and summarize what we observed.

Let's Flip a Coin

■ **Example**. The flipping of a "fair" coin will give us either a head or a tail, with both results being equally probable. That is, approximately 50% of the time we'll get a head, and 50% of the time we'll get a tail. Since **Random[ ]** gives a real number between 0 and 1, we can simulate flipping a coin with:

flip := If[Random[]<0.50, "Head", "Tail"]

It's important to use the colon-equal symbol **:=** to define **flip**. This guarantees that every time *Mathematica* evaluates **flip**, you get a new random number and a new flip of the coin.

flip

Head

flip

Tail

Here are eight coin flips:

Table[flip, {8}]

{Tail, Tail, Head, Tail, Head, Tail, Head, Head}

Count[%, "Head"]   (*Count how many times the head showed up.*)

4

Let's Roll a Die

■ **Example**. A standard die has six sides labeled 1, 2, 3, 4, 5, and 6. If the die is "fair," we expect each of the numbers to appear on the top face about $\frac{1}{6}$ of the time. We can simulate rolling a die with:

Random[Integer,{1,6}]

5

Random[Integer,{1,6}]

1

Now, suppose that players *A* and *B* each roll a die. The one who gets the larger number wins the game. We can simulate this game:

```
game := (   player1 = Random[Integer, {1, 6}];
            player2 = Random[Integer, {1, 6}];
            Which[ player1 > player2, "A wins",
                   player1 == player2, "Tie",
                   player1 < player2, "B wins"] )
```

Each time you evaluate the **game** symbol, you get a new result because we use the **:=** definition.

game

```
B wins
```

game

```
Tie
```

Let us repeat this game 1000 times and count how many times each player wins.

```
data = Table[ game, {1000} ];
Count[ data, "A wins" ]
```

```
393
```

```
Count[ data, "B wins" ]
```

```
421
```

```
Count[ data, "Tie" ]
```

```
186
```

In this case, it seems that player *B* had better luck.

A Strange Behavior of Numerical Answers

■ **Example**. Look at the answer sections in your physics, chemistry, or economics textbooks. Do you notice that the first digit of each numerical answer is more likely to be 1, 2, 3, or 4 rather than 5, 6, 7, 8, or 9? This is because after each multiplication or division of two numbers, the first digit of the result is more likely to be 1, 2, 3, or 4 than the others. (Most scientific values arise from multiplications and divisions.)

This phenomenon is actually related to properties of the logarithm function. If you don't believe it, let's do an experiment to demonstrate it.

We will choose two integers randomly between 1 and 100, multiply them together, and record the first digit of the product.

```
game := ( integer1 = Random[Integer, {1,100}];
          integer2 = Random[Integer, {1,100}];
          First[IntegerDigits[integer1*integer2]])
```

Note that the expression **First[IntegerDigits[ ]]** gives the first digit of the product of the two integers. Here are some test runs:

game

```
1
```

game

```
4
```

Now we repeat this experiment 5000 times. This gives us some experimental data to analyze:

```
data = Table[game,{5000}];
```

The result of "which lead digit occurs most often" is easy to summarize:

```
Needs["Graphics`Graphics`"]
BarChart[ Table[{Count[data, i], i}, {i,1,9}]]
```

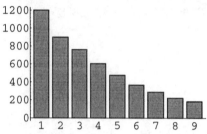

Are you convinced?

A Game of Risk™, Anyone?

■ **Example**. In the Game of Risk™, players attempt to control a map of the world by occupying countries with their armies. In each turn of the game, a player (the "attacker") may choose to engage another player (the "defender") who occupies an adjacent country in battle. If the attacker eliminates the armies of the defender, the attacker takes over the country.

To simulate a battle, the attacker is allowed to roll two dice, but the defender only one. If the attacker's larger die is higher than the defender's die, the attacker wins the battle. However, the defender wins if his die is higher than or ties the larger die of the attacker.

The attacker has the advantage of rolling more dice, but the defender wins ties. So, which player has the advantage?

To find out, we can simulate the action of the attacker by rolling two dice and selecting the largest.

```
attacker := Max[Table[Random[Integer,{1,6}],{2}]]
```

The action of the defender is easier to simulate:

```
defender := Random[Integer,{1,6}]
```

A sample battle might look like this:

```
attacker
```

5

```
defender
```

4

In this case, the attacker wins.

The only outcomes of a battle can be that the attacker wins (+1) or loses (–1). Thus we can simulate a battle with:

```
battle := If[attacker > defender, 1, -1]
```

Now, we can conduct 5000 battles and summarize graphically with:

```
data = Table[battle,{5000}];
Needs["Graphics`Graphics`"]
BarChart[ Table[{Count[data, i], i}, {i,-1, 1, 2}]]
```

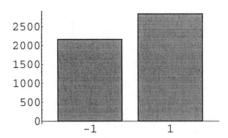

The picture clearly shows that the attacker has the advantage over time. (In fact, the theoretical probability of the attacker winning over the defender is $\left(\frac{5}{6}\right)^3 \approx 58\%$. The graphic shows that our simulation is quite representative.)

Happy Birthday to YOU!!

■ **Example**. The "birthday problem" is a famous problem in probability. For example, in a room of 30 people, how likely is it that at least two of them have the same birthday? We'll suggest an answer using *Mathematica*.

It's sensible to assume that birthdays of 30 randomly assembled people are spread evenly over the year, so we can simulate by picking 30 birthdays at random from 365 days of the year (sorry, we don't do leap years in this experiment!)

```
room := Table[Random[Integer, {1, 365}], {30}]
room
```

```
{66, 97, 185, 58, 148, 212, 105, 280, 164, 289, 301,
    83, 88, 20, 148, 109, 48, 166, 327, 214, 208, 168,
    275, 3, 277, 233, 288, 308, 65, 1}
```

The output above simulates a typical "room" of 30 people, represented by their birthdays. Notice for this output that the number 148 appears twice, indicating that two people of this group had the same birthday on the 148-th day (May 28).

We can count how many *distinct* birthdays there are in the room with:

```
Length[Union[room]]
```

```
29
```

(***Union[room]** sorts the list and *removes duplicates*. **Length[Union[room]]** is then the number of distinct birthdays.*)

When **Length[Union[room]] < 30** is True, there's a duplicate birthday in the room. Now we can experiment with 1000 rooms and count the number of True results we see:

```
data = Table[ Length[Union[room]]<30, {1000} ];
Count[data, True]
```

```
683
```

That is, the probability that there are two people with the same birthday in a room of 30 randomly assembled people is about 683/1000 = 68.3% Are you surprised? Nearly everyone is surprised the first time they see this result! In fact, the theoretical probability that at least two people have the same birthday in a room of 30 people is approximately 70.6%. Our simulation was actually a little lower.

CHAPTER 26
Mathematica for Programmers

This chapter is designed for *Mathematica* users who are already familiar with writing computer programs. However, it is not a tutorial on programming. Rather, we expect that you're looking for information on those areas of *Mathematica* which are closely related to programming concepts.

The first section talks about loops and subroutines, and the second introduces the functional programming capabilities of *Mathematica*. The third section is about pattern matching and recursion. Finally, the fourth section shows you both how to import and export data between *Mathematica* and other computer programs or external data files.

Elements of Traditional Programming Languages

Mathematica has its own programming language that supports elements found in traditional computer programs: conditional execution, looping, and subroutines. The syntax that implements these ideas borrows from the C programming language.

> **Note:** Until now, we used the term "command" to describe *Mathematica* syntax (e.g., we used the "**Plot3D** command" in Chapter 15). In this chapter, most commands will be called "statements." This is more consistent with the terminology of programming languages.

Print Statement

To print the value of one or more expressions, use the statement:

    Print[*expression1* , *expression2* , ...]

The **Print** statement is useful when you are developing programs and are trying to debug them. Some of the examples that follow use the **Print** statement for demonstration.

If Statement

The most-commonly used form of the **If** statement is:

    If[*logical condition* , *result1* , *result2*]

If the *logical condition* evaluates as True, *result1* is returned. If the *logical condition* evaluates as False, *result2* is returned. For example:

    a = 4; If[a <= 3, 2a-1, 5a+1]
    21

    a = 3; If[a <= 3, 2a-1, 5a+1]
    5

Do Loops

The most useful form of the **Do** statement is:

> **Do[** *body* **,** { *variable* **,** *a* **,** *b* } **]**

The *body* of the **Do** statement is executed repeatedly, while the *variable* is automatically incremented from *a* to *b* between iterations. The *body* can be either a single *Mathematica* statement or a sequence of statements separated by semicolons.

■ **Example.** The Fibonacci numbers are $c_0 = 1$, $c_1 = 1$ and $c_n = c_{n-1} + c_{n-2}$, for $n \geq 2$. Executing the assignment **c[n] = c[n–1] + c[n–2]** repeatedly for $n \geq 2$ computes these directly:

```
Clear[c]
c[0] = 1;
c[1] = 1;
Do[ c[n] = c[n-1] + c[n-2], {n, 2, 25} ]
```

To see a few of the Fibonacci numbers, you can use:

```
Table[ c[n], {n, 0, 10} ]
{ 1, 1, 2, 3, 5, 8, 13, 21, 34, 55, 89 }

c[21]
17711
```

For Loops

The general form of the **For** statement is:

> **For[** *initialization statement(s)* **,** *continuation condition* **,**
> *change statement(s)* **,** *body* **]**

For first executes the *initialization statement(s)*. It then repeatedly evaluates the *continuation condition* and, if it is True, executes both the *body* and then the *change statement(s)*. Looping stops as soon as the *continuation condition* becomes False.

Here's a simple example of how the **For** statement works:

```
For[ i=1, i<15, i=i+3, Print[i] ]
1
4
7
10
13
```

> **Note:** Each of the *initialization statement(s)*, *continuation condition*, *change statement(s)*, and *body* in the **For** statement can be either a single *Mathematica* statement or a sequence of statements separated by semicolons.

■ **Example.** To compute the sum of the positive integers 1, 2, 3, …, 50, use:

```
For[ intSum=0; n=1 , n<=50 , n++ , intSum+=n ]
intSum
1275
```

Notice that the two initialization statements **intSum = 0** and **n = 1** are separated by a semicolon. The *change statement* **n++** is a short-hand way of writing **n = n + 1**. The *body* **intSum += n** is a short-hand way of writing **intSum = intSum + n**.

While Loops

The general form of the **While** statement is:

> **While[** *continuation condition* **,** *body* **]**

While repeatedly executes the *body* as long as the *continuation condition* evaluates as `True`. Looping stops once the *continuation condition* becomes `False`.

For example, the second **For** loop given above can be rewritten using a **While** loop:

```
intSum = 0;
n = 1;
While [ n <= 50, intSum += n; n++ ]
intSum

1275
```

> **Note:** Every **For** loop can be rewritten using a **While** sequence and vice versa. Which you use is largely a matter of personal preference.

Procedures, Functions, and Subroutines

Almost every programming language supports a notion of a subroutine, function or procedure. For example, BASIC allows the use of **DEF**, **GOSUB**, and **RETURN** statements, while FORTRAN has a **CALL** statement.

Subroutines are best written in *Mathematica* using the **Module** statement. Its general form is:

> **Module[{** *local variable(s)* **},** *body* **]**

Mathematica will execute the statement(s) in the *body*, then return the value of the last statement executed. Here are two examples.

■ **Example.** To compute the area of a triangle with sides having length a, b and c, you can use Heron's Formula which says that $area = \sqrt{s(s-a)(s-b)(s-c)}$, where $s = (a+b+c)/2$ is the semi-perimeter of the triangle. A nice coding of an area function (subroutine) would be:

```
area[a_,b_,c_] := Module[
        {s} ,
        s = (a+b+c)/2;
        Sqrt[s*(s-a)*(s-b)*(s-c)]
        ]

area[5, 5 ,8]

12
```

> **Note:** Don't forget to add semicolons between statements in the body of a **Module**. Otherwise, *Mathematica* will interpret the space (or line break) between statements as multiplication.

In writing the list of local variables, you can also specify initializations (as you would in C). The **area** routine above could be rewritten as:

```
area[a_,b_,c_] := Module[
        {s = (a+b+c)/2 } ,
        Sqrt[s*(s-a)*(s-b)*(s-c)]
        ]
```

■ **Example.** We can write a routine that computes the maximum, minimum, and average of a list of exam scores from a Calculus class, then reports the scores in increasing order.

```
summarize[data_] := Module[
   { sortedlist, n},
   sortedlist = Sort[data];
   n = Length[data];
   Print["The number of students is ", n];
   Print["The maximum score is ", Last[sortedlist]];
   Print["The minimum score is ", First[sortedlist]];
   Print["The average is ",
         N[Sum[ data[[i]], {i,1,n}]/ n, 4] ];
   sortedlist
   ]

class1 = {61,23,14,78,91,33,12,44,72,79,82,81};
summarize[class1]

The number of students is 12
The maximum score is 91
The minimum score is 12
The average is 55.83
{12, 14, 23, 33, 44, 61, 72, 78, 79, 81, 82, 91}
```

Functional Programming

Mathematica lets you write programs using a *functional* style. If you've programmed using SmallTalk or Lisp, you'll be familiar with the notion of functional programming.

No Loops!   A key element of a functional programming style is the absence of explicit looping. Rather, you use available statements and functions that construct and manipulate entire lists. You also use expression transformations and operator compositions.

■ **Example.** To compute the sum of the entries of a list of numbers named **myList**, we previously suggested the coding:

```
myList = { 3, 5, 8, 10, 2, 6 };
Sum[ myList[[i]], {i,1,Length[myList]} ]

34
```

Using a functional programming style, we can take advantage of the fact that every expression and every object in *Mathematica* has an internal, structured form that consists of its *head* and its *arguments*.

The form of **myList** as an object is **List[3, 5, 8, 10, 2, 6]**. You can find this out using the **FullForm** command:

```
FullForm[myList]

List[3,5,8,10,2,6]
```

To get the sum of the elements in this list, we need to form the expression $3 + 5 + 8 + 10 + 2 + 6$. This expression also has an internal form in *Mathematica* that is **Plus[3, 5, 8, 10, 2, 6]**.

Expression	*Internal Form*
{ 3, 5, 8, 10, 2, 6 }	List[3, 5, 8, 10, 2, 6]
3 + 5 + 8 + 10 + 2 + 6	Plus[3, 5, 8, 10, 2, 6]

To sum up the entries in the list **myList**, all you need to do is have *Mathematica* change the *head* of the expression from **List** to **Plus**. This is exactly what happens when you use the **Apply** statement:

```
Apply[Plus,myList]
   34
```

This use of **Apply** is both powerful and extremely efficient. Even on a short list such as the one above, the execution timings of the two ways to sum the entries of the list differ by a factor of about 10.

Functional Iteration

■ **Example.** To estimate the square root of a positive number A, you can iterate the expression $f(x) = \frac{1}{2}(x + A/x)$. To approximate $\sqrt{3}$, say, you could use a loop that runs four times, starting at $x_0 = 2.0$:

```
x = 2.0;
For[ i=1 , i<=4 , i++ , x = 1/2 (x+3/x); Print[x] ]
   1.75
   1.73214
   1.73205
   1.73205
```

With a functional approach, you use the **NestList** statement. Its general form is:

> **NestList[** *an iteration function f* **,**
> 　　　　　　　　 *a starting value* x_0 **,** *an iteration count n* **]**

NestList iterates the function f exactly n times starting from the value x_0 to produce the list having elements:

$$\{x_0, x_1 = f(x_0), x_2 = f(x_1), \dots, x_n = f(x_{n-1})\}.$$

The functional equivalent for approximation of $\sqrt{3}$ given above is written:

```
Clear[f,x]
f[x_] := 1/2 (x+3/x)
NestList[ f, 2.0, 4 ]
   {2., 1.75, 1.73214, 1.73205, 1.73205 }
```

The Functional Programming Style

■ **Example.** Define a positive integer to be *perfect* if it is the sum of its smaller divisors. For instance, both 6 and 28 are perfect since $6 = 1 + 2 + 3$ and $28 = 1 + 2 + 4 + 7 + 14$.

We'll write functional coding to find all perfect numbers up to 1000. The nicely demonstrates the functional style of programming.

The idea is first to write a function that tests whether a given integer n is perfect. We can use the **Divisors** command to form a list of all the divisors of n and then add up all but n itself. The number n is perfect if this sum equals n:

```
isperfect[n_] := Module[ {properDivisors},
        properDivisors = Drop[Divisors[n],-1];
        Apply[Plus,properDivisors]==n
        ]
```

Once this test is available, the functional construct **Select** can be used to choose from among the integers 1, 2, ... , 1000 those that are perfect:

```
Select[Range[1,1000],isperfect]
```

{6, 28, 496}

Pattern Matching

Pattern Matching in Defining Functions

Although *Mathematica* may seem to be a knowledgeable mathematics program, it is actually more of a text-processing system at its core. You write expressions as input, and *Mathematica* rewrites and simplifies them as best as it can to report its output.

For example, you may think of the *function* $f(x) = x^2$ as a mathematical object defined by writing:

```
f[x_] := x^2
```

Mathematica actually records your definition as a way to rewrite expressions as:

whenever you see the expression "f(something)," rewrite it as "(something)^2."

The definition of **f** is called a *rule*. It tells *Mathematica* how to rewrite an expression that looks like **f**[*something*].

Changing the Rules

Because *Mathematica* treats expressions under very general assumptions, certain simplifications we expect to occur may not happen. For example, the **Sqrt** function does not treat $\sqrt{x^2}$ as x since this is only sensible if x is a non-negative, real number:

```
Clear[x]
Sqrt[x^2]
```

$$\sqrt{x^2}$$

We can use pattern matching to write our own version of a square root function that simplifies $\sqrt{x^2}$ to be x with:

```
Clear[mySqrt]
mySqrt[x_^2] := x
```

The variable above is the pattern **x_^2**. *Mathematica* will successfully match this pattern against any expression which has the form **mySqrt**[*something^2*]. In other words, **mySqrt** has a rule associated with it which says:

*whenever you see the expression "**mySqrt**[something^2],"*
rewrite it to be just the "something."

Now let's see some examples:

```
Clear[a,b]
mySqrt[a^2]
```

a

```
mySqrt[(a+b+1)^2]
```

a+b+1

However, we haven't said anything about how **mySqrt** should handle things which are not squares, as you can see with:

mySqrt[4]

mySqrt[4]   (*Mathematica returns the usual answer because **4** is
                   not of the form "something^2".*)

To fix this deficiency and make **mySqrt** act just like **Sqrt** for cases like the one above, we introduce a second rule for its definition:

mySqrt[x_] := Sqrt[x]

This second rule does not conflict with the first. *Mathematica* will check against both of the patterns we've defined to figure out how to simplify expressions:

mySqrt[4]   (*Mathematica uses the **mySqrt[x_]** pattern.*)

2

mySqrt[a^2]   (*Mathematica uses the **mySqrt[x_^2]** pattern.*)

a

Conditional Pattern Matching

Our definition of a new square root function still needs some work because if we try:

mySqrt[(a+b)^4]

$$\sqrt{(a+b)^4}$$

Mathematica returns the answer unevaluated because the expression **(a+b)^4** is not technically of the form "*something^2.*" To handle this case, you must use what's called a **conditional pattern matching rule**. It is invoked by *Mathematica* only when the power is an even integer:

```
Clear[mySqrt]
mySqrt[x_] := Sqrt[x]
mySqrt[x_^n_Integer?EvenQ] := x^(n/2)
```

Now we test our definition:

mySqrt[(a+b)^4]

$$(a+b)^2$$

mySqrt[(1-a-b)^(-6)]

$$\frac{1}{(1-a-b)^3}$$

mySqrt[x^3]   (*The power 3 is not even; no simplification is done.*)

$$\sqrt{x^3}$$

Pattern Matching in Substitution

Pattern matching is used to do substitution. For example, the following statement asks *Mathematica* to locate the pattern **a**, then replace it with **b^2**.

    a + 2*c /. {a -> b^2}   (*Substitute **b^2** for **a**.*)

    $b^2 + 2\,c$

Similarly, if you want to replace the entire expression $\sin(2x)$ by $2\sin(x)\cos(x)$ in an expression, you could use this substitution rule:

    5a+Sin[2*x] /. {Sin[2*x] -> 2*Sin[x]*Cos[x]}

    $5\,a + 2\,\mathrm{Cos}[x]\,\mathrm{Sin}[x]$

Mathematica will look for the pattern **Sin[2*x]**, and replace it with the given expression. However, if you try:

    5a+Sin[2*x]+Sin[2*y] /. {Sin[2*x] -> 2*Sin[x]*Cos[x]}

    $5\,a + 2\,\mathrm{Cos}[x]\,\mathrm{Sin}[x] + \mathrm{Sin}[2\,y]$

The expression **Sin[2*y]** is not replaced because it does not match the specified pattern **Sin[2*x]**. If you want to change this one too, you must use a pattern variable such as **x_** (rather than a literal **x**) in the substitution rule:

    5a+Sin[2*x] +Sin[2*y] /. {Sin[2*x_]->2*Sin[x]*Cos[x]}

    $5\,a + 2\,\mathrm{Cos}[x]\,\mathrm{Sin}[x] + 2\,\mathrm{Cos}[y]\,\mathrm{Sin}[y]$

Mathematica will look for any pattern of the form "**Sin[2****something***]**" and then replace according using the given replacement rule. In this case, both **Sin[2*x]** and **Sin[2*y]** will be replaced.

Finally, since *Mathematica* doesn't simplify $\ln(e^x)$ to be x (even the **FullSimplify** command in version 3.0 won't do it for you), we can force this to be done through pattern matching:

    Log[E^x]+Log[E^(y^2)]+Log[E^(3z)] /. {Log[E^x_]->x}

    $x + y^2 + 3\,z$

Recursion

Because *Mathematica* repeatedly expands expressions using pattern matching until they no longer change, recursive computations are done quite naturally.

For example, the Fibonacci numbers are defined recursively as $c_0 = 1$, $c_1 = 1$ and $c_n = c_{n-1} + c_{n-2}$, for $n \geq 2$. These can be defined in *Mathematica* with:

```
Clear[c]
c[0] = 1; c[1] = 1;
c[n_] := c[n-1] + c[n-2]
```

The first two definitions for **c[0]** and **c[1]** are specialized cases for pattern-matching. The more general definition for **c[n_]** is used for expressions that are of the form **c[** *something* **]**. To find the Fibonacci number c_5, just type:

    c[5]

    8

To evaluate **c[5]**, *Mathematica* repeatedly uses the three rules above to simplify the expression in approximately the following sequence of transitions:

$$c_5 \rightarrow c_4 + c_3 \rightarrow (c_3 + c_2) + (c_2 + c_1) \rightarrow ((c_2 + c_1) + (c_1 + c_0)) + ((c_1 + c_0) + 1)$$

$$\rightarrow (((c_1 + c_0) + 1) + (1 + 1)) + ((1 + 1) + 1) \rightarrow (((1 + 1) + 1) + 2) + (3) \rightarrow 8$$

Remembering Values During Recursion

One disadvantage of the method above is that *Mathematica* will not remember the values that it has calculated recursively. For example, if you want to calculate $c[6]$, *Mathematica* recomputes $c[5]$, $c[4]$, $c[3]$, and $c[2]$ all over again. This can slow down computation dramatically, even for just, say, $c[20]$.

Mathematica's pattern-matching mechanism gives you a way to evaluate $c[5]$ and at the same time define its value in case you need to use it later You do this with the following pattern definition:

```
Clear[c]
c[0] = 1; c[1] = 1;
c[n_] := c[n] = c[n-1] + c[n-2]
```

The pattern on the right side of the delayed assignment := is now an assignment statement.

Evaluation of $c[5]$ causes *Mathematica* to execute the assignment statement $c[5] = c[4] + c[3]$, which means that the computation is done *and the result is assigned to* $c[5]$. Now, next time if you want to compute, say, $c[7]$, *Mathematica* only needs to compute $c[6]$ and can then give you the answer.

The effect on execution time is dramatic. Evaluating $c[20]$ using the first recursive method above takes *two to three hundred* times longer than the second definition for $c[20]$ which remembers its values!

File I/O

Saving Values to a File

If you want to save (numerical) values you compute in *Mathematica* to a file, you usually follow a three-step process.

- First, you must create an output channel reference for the output file. You use the **OpenWrite** statement to do this, naming the file to which you will write the data:

```
channel = OpenWrite[" a valid file name for your system "]
```

- Next, you use the **Write** statement to output the values. You may be able to write all of the values in a single **Write** statement or you may use a loop to write the values one at a time.

The general form of the statement is:

```
Write[ channel, value1 , value2 , etc. ]
```

Each of the values you list is written to the file and separated by a space. A newline character is written to the file after the **Write** is performed.

- Finally, when you are finished with output, you should **Close** the file:

```
Close[channel]
```

■ **Example.** Suppose that you want to compute the values of the Fibonacci numbers and then save those values to a file named "**Fibonacci.dat**". You'd do this with the following sequence (which writes 25 numbers for demonstration):

```
Clear[c];  c[0] = 1;  c[1] = 1;
c[n_] := c[n] = c[n-1] +c[n-2]      (*This defines the numbers.*)

channel = OpenWrite["Fibonacci.dat"]
Do[ Write[channel,c[n]], {n, 0, 25} ]
Close[channel]
```

Importing Numerical Values from a File

To import data values from another application (e.g., a spreadsheet or an e-mail message), you should use the **OpenRead** and **ReadList** statements. **OpenRead** will create an input channel reference. **ReadList** will read all the data values into a *Mathematica* list that you can then process as you wish.

Suppose, for example, that you want to read all the data from a file **TestScores.dat** into a list named **dataValues**. You type:

```
channel = OpenRead["TestScores.dat"];
dataValues = ReadList[channel];
Close[channel];
```

Now you can work with the data quite easily. For example, the mean of the data is simply:

```
Needs["Statistics`Master`"]
Mean[dataValues]
```

If you have limited memory and a large data set, it may not be possible to have all of the data values stored in memory at the same time in a single list. In that case, you can read and process one value at a time using the **Read** statement.

For example, the following loop provides a skeleton structure to process the values in **TestScores.dat**. It repeatedly executes **Read** to input a single data value and prints it. The loop ends when **Read** returns the special value **EndOfFile** to signal that no more data is available in the file.

```
channel = OpenRead["TestScores.dat"]
While[True,
       oneDataValue = Read[channel];
       If[ oneDataValue==EndOfFile, Break[] ];
       Print[oneDataVlue]
       ];
Close[channel];
```

APPENDIX
Working with Notebooks

Front End and Kernel

Beginning with version 2.1, every graphically-based system of *Mathematica* has been split into components called the **Front End** and the **Kernel**. Each of these is itself an application.

The Kernel

The *Mathematica* **Kernel** is an application that does all the computational work. For example, when you solve a system of equations, it's the **Kernel** software that interprets the symbols you use and computes a solution in terms of the variables.

You usually don't work with the **Kernel** directly.

The Front End

The *Mathematica* **Front End** is an application that gives you a textual and graphical interface to the **Kernel**. When you start *Mathematica*, you are working with the **Front End**. After you enter an expression and ask for its evaluation, the **Front End** automatically starts up the **Kernel**, passes the input to it, and displays the result.

For example, when you input:

```
Integrate[ 6x^2, {x, 0, 1} ]
```

the **Front End** passes this expression to the **Kernel**. It is there that integration is performed and the result of 2 is computed. This result is then passed back to the **Front End** and displayed:

2

Notebooks and Cells

Notebooks

During each *Mathematica* session, you can save the record of your inputs and outputs. This computer file you saved is called a **Notebook**. Notebooks don't carry out any computation – they only provide a record of the traffic that goes back and forth between the **Front End** and the **Kernel**. In other words, Notebooks are static text documents.

Cells

Every notebook you work with in the **Front End** is composed of **cells**. Every *Mathematica* input you make will occupy an "input cell," and every output you receive from *Mathematica* will occupy an "output cell." These cells are indicated by "cell brackets" that appear on the right side of the notebook window, as shown here:

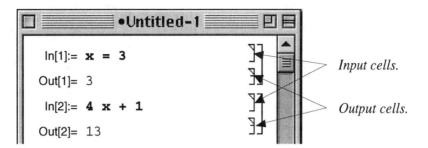

Input cells.

Output cells.

Cell Type

Every cell has a "type" associated with it, depending on its intended usage. By default, when you start typing at a new location in a Notebook, you are creating an *input* cell. When you evaluate your input, the result is automatically in an *output* cell.

In addition to the input and output cells, there are other styles such as text, title, section, and subsection. Let's show you how to create a text cell.

Say, you want to add a commentary before the input **4x+1**. To do so, follow the instructions below:

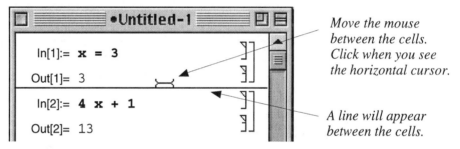

Move the mouse between the cells. Click when you see the horizontal cursor.

A line will appear between the cells.

You can start typing your commentary:

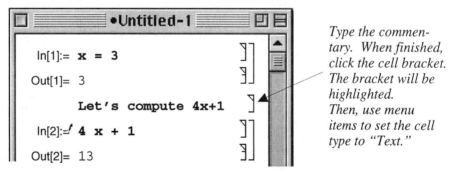

Type the commentary. When finished, click the cell bracket. The bracket will be highlighted. Then, use menu items to set the cell type to "Text."

(If you are using version 2.x, go to the Style menu on the menu bar, open "Cell style," and choose "Text." If you are using version 3.0, go to the Format menu on the menu bar, open "Style," and choose "text".) The cell will change immediately into text style.

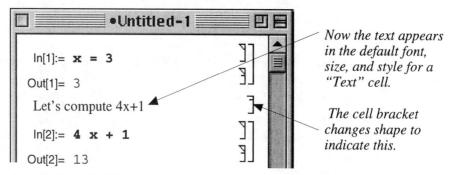

Now the text appears in the default font, size, and style for a "Text" cell.

The cell bracket changes shape to indicate this.

Using a similar procedure, you can also change the cell style to title, subtitle, message, and so on. Each of these styles has its own format and serves different purposes. You can experiment with them.

Editing Cells and Text

The following table summarizes the editing procedures you'll use the most when working with cells:

What You Want to Do	*How to Do It*
Start a new cell in a notebook.	• Move the mouse to the new location which is between or outside existing cells. Wait for the cursor to change to the horizontal insertion shape ⊢⊣. • Click the mouse. A horizontal line will appear between cells. • Start typing.
Delete a cell.	• Click the cell bracket to select the cell. The bracket will be highlighted. • Choose either the **Cut** or **Clear** command from the Edit menu.
Make a copy of a cell in a new location.	• Click the cell bracket to select the cell. The bracket will be highlighted. • Choose the **Copy** command from the Edit menu. • Move the mouse to the new location. Wait for the cursor to change to the horizontal insertion shape ⊢⊣. Click the mouse. A horizontal line will appear between cells. • Choose the **Paste** command from the Edit menu.
Move a cell to a new location.	• Click the cell bracket to select the cell. The bracket will be highlighted. • Choose the **Cut** command from the Edit menu. • Move the mouse to the location where you want the cell to appear. Wait for the cursor to change to the horizontal insertion shape ⊢⊣. Click the mouse. A horizontal line will appear between cells. • Choose the **Paste** command from the Edit menu.

Cut, copy, or paste the text of a cell within the same or another cell.	• Handle this the same way that you manipulate text in any word processor. (Use the mouse to select, and then use one of the **Cut**, **Copy** or **Paste** commands.)
Change the font, size, or style of an entire cell.	• Click the cell bracket to select the cell. • From either the **Style** menu (for version 2.x) or **Format** menu (for version 3), select the appropriate font, size, and style.
Change the font, size, or style of some (or all) of the text within a cell.	• Select the text with the mouse. • From either the **Style** menu (for version 2.x) or **Format** menu (for version 3), select the appropriate font, size, and style.
Change the default font, size, or style of all the cells	• Refer to your *Mathematica* user manual, because various versions have different procedures.

Notebook Structure

Cell Type and Notebook Organization

A well-organized document has a title, perhaps a subtitle, and is divided into a number of sections. The sections may further be broken into subsections.

You can structure a *Mathematica* notebook in much the same way by using a title cell, section cells, and subsection cells. Using these together with input and output cells yields a nicely organized document that's suitable both for reading and printing.

■ **Example.** Suppose we have made the following simple calculus computations in a notebook:

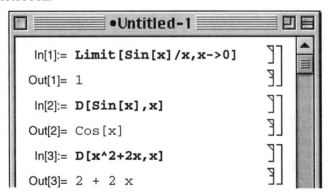

To organize these results, we will add a section heading to say that these are from "Calculus", and then add two subsection headings that separate them into "Limits" and "Derivatives". We do this by first adding three cells. (Make sure you review the instructions in the previous section on how to add cells and change their type.)

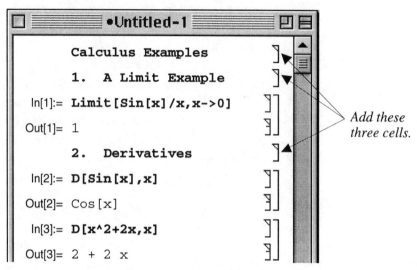

For each of the newly-added cells, click the cell bracket at the right and change the type of the cell to "Section" and "Subsection." The cells will be reformatted for their new type.

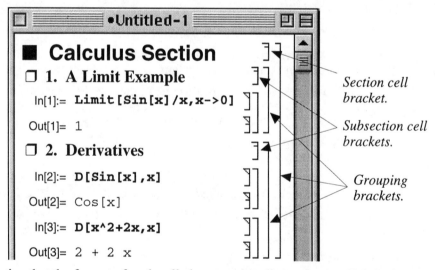

Notice that the format of each cell changes according to how cell defaults are set up and installed on your machine. The Section and Subsection cell types will also be shown with leading characters called *dingbats* to distinguish them (you don't type these characters – they're added as part of the cell format).

Cell Grouping   New bracketing has been added automatically at the right of the notebook window to reflect its organization into one Section, two Subsections and a number of input and output cells. (See the picture above.)

The outer-most bracketing shows that all cells are grouped and headed by the Section cell. The next outer-most bracketing shows two groups, each headed by Subsection cells. All output cells are grouped with their associated input cell.

Open and Closed Cell Groups

All the cell groups of the previous window are said to be *open*, because you can see their contents. You can *close* cell groupings so that you only see the predominant cell in each group. This provides a very nice outlining capability, as you'll see.

For example, the "Derivatives" subsection can be closed by double-clicking its grouping bracket.

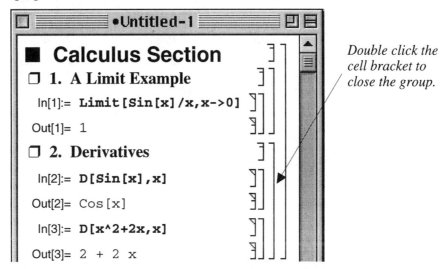

Double click the cell bracket to close the group.

With the group closed, the notebook window now looks like this:

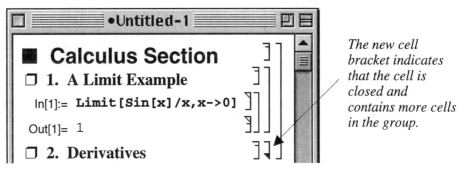

The new cell bracket indicates that the cell is closed and contains more cells in the group.

If you also close the "Limit Example" subsection, the window becomes even more compact.

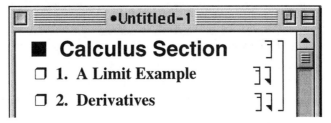

To open cells, double-click on the grouping brackets.

Input Shortcuts for Version 3.0

In *Mathematica* 3.0 (or higher), you can find the "Palettes" command in the "File" menu in the Front End. Palettes are special types of notebooks which provide short-cuts to entering commands and expressions. They're especially useful for preparing notebooks for printing and publication.

There are seven predefined palettes. We'll quickly illustrate how three of them work. You'll want to experiment with all of them, as well as consult your software guide to learn more about them.

The Algebraic Manipulation Palette

This palette gives quick and easy access to commands such as **Expand**, **Simplify**, **Apart**, and so on. What's interesting is that you can apply commands from this palette to a selection anywhere in a notebook.

For example, suppose you input the following polynomial:

(x-2)(x+3)^3

$$(-2 + x) \ (3 + x)^3$$

You now decide to **Expand** the result. Select the "AlgebraicManipulation" command from the "Palettes" submenu that's located in the "File menu." The palette will open up on screen in its own window, outside your Notebook window. (A portion of it is shown to the right.)

Now select the output (either select the entire output result with the mouse, or select the output cell itself by clicking on its bracket), and click the Expand[■] button in the palette. *Mathematica* copies the selected expression to a new input cell, briefly shows you the expression **Expand**[$(-2 + x) \ (3 + x)^3$], and then expands it *in place* to rewrite the input cell as:

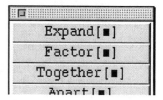

$$-54 - 27 \ x + 9 \ x^2 + 7 \ x^3 + x^4$$

The Basic Calculations Palette

This palette provides templates for various command inputs. It is organized into categories and subcategories so that you can locate commands easily. Clicking on the triangles which appear next to the category names rotates the triangle and opens up the list of subcategories.

■ **Example**. To solve the equation $3x + 1 = 5$, open the **Basic Calculations** palette. Under the "Algebra" section, choose "Solving Equations." (See the picture on the right.) Now, click `Solve[■ == □, □]`. A new input cell will be created in your notebook:

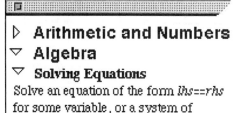

  `Solve[ ■ == □, □]`

Now type **3x+1**, then press the *tab* key. Next, type the **5**, then press the *tab* key again. Finally, type the **x**. (Every time you press the *tab* key, you move from one open box in the expression to the next box.) Now, you can execute the command.

  `Solve[3x+1==5,x]`

$$\left\{ \left\{ x \rightarrow \frac{4}{3} \right\} \right\}$$

The Basic Input Palette

The Basic Input palette lets you enter expressions involving integrals, roots, and fraction in a more pleasing, mathematical way. It also provides buttons to enter Greek symbols and special characters directly.

■ **Example**. In version 2.2, you could only evaluate $\int x^2 \, dx$ by typing:

  `Integrate[ x^2, x ]`

In version 3.0, you can choose the **BasicInput** palette and click on the ⌠■d□ button. A new input cell is created in your notebook that looks like:

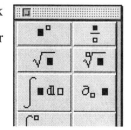

$$\int ■ \, d\square$$

Next, click on the ■□ button. Now you see:

$$\int ■^\square \, d\square$$

Next, type *x*, then *tab*, then 2, then *tab*, and then *x* again. Every time you press the *tab* key, you move from one box in the expression to another box. Now you see the completed expression:

$$\int x^2 \, d \, x$$

Evaluating this input cell, you now get the expected result:

$$\frac{x^3}{3}$$

Greek Letters and the Basic Input Palette

Prior to version 3.0 of *Mathematica*, you couldn't really write Greek! You might define a variable named **alpha**, but the effect wasn't as comforting as seeing the Greek character α directly in front of you.

In version 3.0, however, you can enter the Greek character α directly into a notebook from the **BasicInput** palette by clicking the $\boxed{\alpha}$ button. You'll find buttons for all the Greek letters in a lower section of the Basic Input palette. Clicking on one of them types the character for you in your notebook.

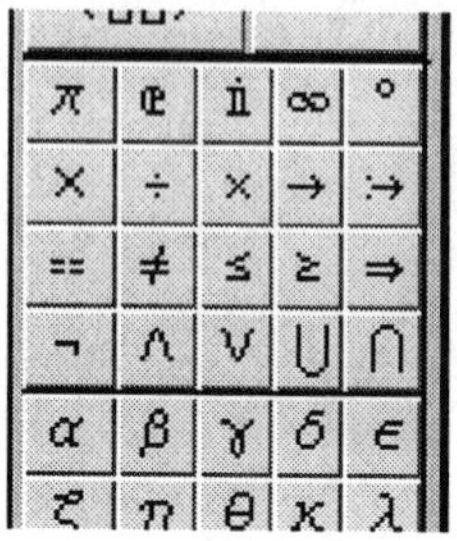

For example, you can input an expression which assigns 3 to the variable α by clicking the $\boxed{\alpha}$ button, then typing "=" and "3":

> α = 3
>
> 3

Now you can use the symbol α anywhere you'd use a variable, such as with:

> **4 α + 1**
>
> 13

Just as easily, you don't have to type **Pi** to represent the constant π. You can choose it directly from the palette:

> **π // N**
>
> 3.14159

It now becomes very easy to write meaningful *Mathematica* to see, say, the unit sphere in 3-D drawn parametrically using spherical coordinates. (The output is not shown here, however.)

```
ParametricPlot3D[{Cos[θ]Sin[φ],Sin[θ]Sin[φ],Cos[φ]},
                 {θ,0,2π}, {φ,0,π}]
```

> **Note:** You can use the Basic Input palette to enter an expression anywhere in a notebook – even within a text cell. You'll want to do this if you're creating a notebook that you want to be printed using proper mathematical notation.

Index

=, 18
^, 8
!, 3
%, 9
(* and *), 9
*, 8, 39
+, 8
++, 161
+=, 161
−, 8
/, 8
/., 16
<<, 42
?, 38
??, 38
(*parentheses*), 3, 8, 30,
[*square brackets*], 3, 11, 17, 30
[[*double square brackets*]], 29, 30, 109
{ *curly braces* }, 3, 17, 30
:=, 17
;, 48, 147
==, 19, 26
>, 19
>=, 19
<, 19
<=, 19
!=, 19
&&, 19
| |, 19
_, 17, 18
->, 17

2-D parametric curve 51-54
2-D plotting, 43-49
3-D parametric curve, 114-115
3-D plotting, 85-93

Absolute value, 11
Abs, 11
Add, 8
algebra, 4, 21-26
All, 44
Animate, 140
animation, 140-146
antiderivative, 67, 70
Apart, 23
Apply, 164
approximate numeric values, 10, 12

area between curves, 69
arithmetic, 8
aspect ratio, 44
AspectRatio, 37, 44, 51, 56, 59
Automatic, 37, 44, 51, 56, 112, 119
Axes, 86, 112, 119
AxesLabel, 35, 45, 86

BarChart, 126, 152, 155
BASIC, 162
beehive, 114
Bessel function, 6
BesselJ, 6
BetaDistribution, 128
BinomialDistribution, 128
birthday problem, 159
browser-based help, 39
boundary conditions, 80
Boxed, 86, 112, 114
BoxRatios, 86, 92, 112
Break, 169
Built-in Constants, 10
Built-in Functions, 11

C programming, 5, 160,
calculator, 8
Calculus, 60-72
Calculus`FourierTransform`, 75
Calculus`VectorAnalysis`, 121
cardioids, 56
Cartesian, 121
CDF, 129
cell bracket, 171
cell insertion point, 171
cell styles, 171
cells, 170
central tendency, 128
chain rule, 63
ChiDistribution, 128
Chinese character for "Wind", 57
ChiSquareDistribution, 128
Clear, 16, 19
Clear["@"], 20
Close, 131, 168
color, 47, 57, 115
ColumnForm, 147
comments, 9
common denominator, 22

compilation, 54
complex numbers, 21
confidence interval, 128
ConfidenceLevel, 128
ContourPlot, 94
ContourPlot3D, 96
Contours, 94, 96
ContourShading, 94, 123
Count, 148
critical points, 102
Cross, 106
cross product, 106
Cumulative distribution function, 129
Curl, 121
curve fitting, 133-139
cycloid, 52, 143
Cylindrical, 121
cylindrical coordinate system, 89, 121
CylindricalPlot3D, 89

D, 4, 62, 100
daisy, 5
damped oscillation, 144
Dashing, 47
data massaging, 149
data transformation, 149
data variability, 128
definite integral, 68
Degree, 11
delayed assignment, 18, 20
derivative, 4, 61
DescriptiveStatistics, 41
Det, 108
determinant, 108
differential equations, 79-84
differentiation, 62, 100
Direction, 61
DiscreteMath`Permutations`, 152
discriminant, 102
DisplayFunction, 142
Div, 121
Divide, 8
Divisors, 164
Do, 161
DOS, 1
Dot, 106, 108
dot product, 106
double integral, 100
Drop, 149, 150
DSolve, 79, 81

E, 10
Eigenvalues, 108
Eigenvectors, 109
ellipsoid, 98

EndOfFile, 169
equation, 26
Erf, 71
error message, 12
Evaluate, 66, 75
evaluation key, 2
EvenQ, 166
Expand, 4, 21, 23
Exponential, 10, 11
ExponentialDistribution, 128

FaceForm, 115
Factor, 4, 21, 23
factorial, 3
Fibonacci numbers, 161
FilledPlot, 69
FindRoot, 3, 28, 29, 38
First, 128, 149
Fit, 133
flipping of a coin, 5, 156
For, 161
FORTRAN, 162
FourierSeries, 75
FourierTrigSeries, 75
frame function, 142
Frames, 142
FresnelC, 71
FresnelS, 71
Front End, 170
FullForm, 163
FullSimplify, 24, 167
Function Browser, 39
function defining, 7, 18, 19,
functional programming, 163
Fundamental Theorem of Calculus, 68, 70

Game of Risk™, 158
GeometricMean, 128
GoldenRatio, 57, 59
Grad, 121
gradient, 121
Graphics, 41
Graphics`Animation`, 140
Graphics`ContourPlot3D`, 96
Graphics`FilledPlot`, 69
Graphics`Graphics`, 41, 55, 58, 126, 152, 155
Graphics`ImplicitPlot`, 56, 59
Graphics`Master`, 124
Graphics`ParametricPlot3D`, 89
Graphics`PlotField3D`, 120
Graphics`PlotField`, 83, 119
graphing function, 43-50
GrayLevel, 52, 56, 115
grouping bracket, 174

HarmonicMean, 128
Heaviside function, 18
helix, 112
Help Browser, 39
Help command, 38
Heron's Formula, 162
Hessian Test, 102
hyperbolic functions, 11, 23
hyperboloid, 98, 113

I, 10
Identity, 142
identity matrix, 108
IdentityMatrix, 108
If, 6, 156, 160
Imaginary number, 10
immediate assignment, 15, 20
ImplicitPlot, 56, 59, 97
import data, 131
improper integral, 68
infinite series, 72, 73
Infinity, 11
initial condition, 80
input cell, 170, 171
input style, 171
Integer, 166
IntegerDigits, 157
Integrate, 4, 67, 100, 122
integration, 67-71, 100, 122
InterpolatingFunction, 80
InterpolatingPolynomial, 135
interpolation, 135
Intersection, 150
Inverse, 108, 111
Inverse hyperbolic functions, 11
Inverse trigonometric functions, 11
IQ scores, 129

Kernel, 170

Lagrange Multiplers, Method of, 103
Last, 149
least square method, 133
left-hand limit, 61
Length, 149, 150
level curves, 94
level surfaces, 96
Lighting, 116
Limit, 60
limit calculation, 60, 61
line integral, 122
linear regression, 133
LinearAlgebra`CrossProduct`, 106
LinearAlgebra`Master`, 107
LISP, 163

list, 33
ListPlot, 35, 53, 126
loading packages, 41
local variable, 162
logarithm, 11
logistic model, 137
long division, 23

MacOS, 1
Map, 128, 132, 138, 148
Mathematica Book, 40
matrices, 107
MatrixForm, 107
MatrixPower, 108
Max, 149
Mean, 41, 42, 128, 150
MeanCI, 128
MeanDeviation, 128
Median, 41, 128
memory, 7, 92
Min, 149
Module, 6, 162
Multiply, 8

N, 10, 12
name, 15, 16
NDSolve, 80, 81
Needs, 41
NestList, 164
network, 1
NIntegrate, 68, 70, 102
nonagon, 36
Normal, 74
normal distribution, 128
NormalDistribution, 128
Notebook, 170
NSolve, 28,
NSum, 72
numerical computation, 10, 12
numerical differential equation, 80
numerical integration, 68, 70, 102

OpenRead, 131, 168
OpenWrite, 168
output cell, 171
output style, 171

packages, 41
Palettes, 176
paraboloid, 114
ParametricPlot, 5, 51
ParametricPlot3D, 5, 112, 113
parentheses, 8
partial differentiation, 100
partial fractions, 23

pattern matching, 165-168
PDF, 129
perfect numbers, 164
perpendicular property, 123
phase diagram, 82
Pi, 10
PieChart, 41, 126
Plot, 4, 43, 50, 66
Plot3D, 5, 85
PlotJoined, 35
PlotLabel, 45
PlotPoints, 86, 94, 119
PlotRange, 43, 86
PlotStyle, 35, 46, 56, 115
PlotVectorField, 83, 119
PlotVectorField3D, 120
PointSize, 35
PoissonDistribution, 128
polar coordinates, 55, 104
PolarPlot, 55
polynomials, 21
postfix syntax, 10, 22, 107, 147
power, 8
PowerExpand, 22
precedence, 9
Prime, 148
Print, 160
probability density function, 129
product rule, 62
programming language, 5, 160, 163
pseudo-random, 154

quadratic formula, 17
Quadric Surfaces, 98
Quit, 2
quotient rule, 63

radian measure, 12
Random, 130, 154
RandomPermutation, 152
random numbers, 154
Range, 165
rational function, 22
ReadList, 131
RealDigits, 151
recursion, 167
Regress, 134
regression, 135
Remove, 42, 58, 59, 83, 99, 131, 146
return key, 6
Reverse, 132, 150, 151
RGBColor, 47, 57, 115
right-hand limit, 61
roll a die, 56

RootMeanSquare, 128
row echelon form, 110
row transformations, 109
row vectors, 109
RSquared, 134

saddle, 85
saddle point, 103
scalar multiplication, 106
ScaleFunction, 83, 119, 124
scientific notation, 10
section cell, 173
Select, 165
semicolon, 48, 147
Series, 74
Shading, 115
Shallow, 148
Short, 148
Show, 37, 46, 50, 53
Simplify, 4, 21, 22, 23
simulation, 155-159
SmallTalk, 163
solution curve, 82
Solve, 3, 26, 27
sombrero, 86
Sort, 150
special functions, 71
Spherical, 121
spherical coordinate system, 89, 90, 121
SphericalPlot3D, 89
split definition, 76
square root, 11
standard packages, 41
StandardDeviation, 128
Statistics`DescriptiveStatistics`, 41
Statistics`Master`, 127, 131
string, 33
StudentTDistribution, 128
style directives, 115
subsection cell, 173
substitution rule, 17
substitution symbol, 16
subtract, 8
Sum, 72, 150
surface, 85, 113
surface integral, 122, 123
system of equations, 27

Table, 34
TableForm, 147, 151
Take, 149
TaylorPolynomials, 73
Taylor Series, 73
text cell, 171
text editing, 172, 173

text style, 171
title cell, 171, 173
Thickness, 46, 115
three-leaf rose, 55
Together, 22
Trig, 23
trigonometric functions, 11, 23
trigonometric identities, 4, 23
triple integral, 102
True, 26

uniform distribution, 155
UniformDistribution, 128
Union, 150
UNIX, 1
U.S. population model, 137

variable names, 15, 16
Variance, 128
VarianceCI, 128
vectors, 106
vector field, 119
ViewPoint, 87, 92

Which, 18, 76
While, 6, 162
wild-card character, 39
Write, 168

Zooming, 48